Texts in Computer Science

Series Editors

David Gries, Department of Computer Science, Cornell University, Ithaca, NY, USA

Orit Hazzan, Faculty of Education in Technology and Science, Technion—Israel Institute of Technology, Haifa, Israel

More information about this series at http://www.springer.com/series/3191

Dengsheng Zhang

Fundamentals of Image Data Mining

Analysis, Features, Classification and Retrieval

 Springer

Dengsheng Zhang
Federation University Australia
Churchill, VIC, Australia

ISSN 1868-0941   ISSN 1868-095X   (electronic)
Texts in Computer Science
ISBN 978-3-030-17991-5   ISBN 978-3-030-17989-2   (eBook)
https://doi.org/10.1007/978-3-030-17989-2

This Springer imprint is published by the registered company Springer Nature Switzerland AG.
The registered company address is: Gewerbestrasse 11, 6330 Cham, Switzerland

To my beloved wife Qin who makes this possible

Preface

Due to the rapid development of Internet and digital technology, a mammoth amount of digital data has been created in the world in just a few decades. The processing and management of big data including image data have become one of the great challenges facing humankind. Images are a dominant information source and communication method along with text. However, the processing and understanding of image data are far more difficult than dealing with textual data. Tremendous efforts have been made, and a large amount of research work has been carried out around the world in the past two decades to overcome the challenges of efficient management of image data. A significant progress has been achieved in the field of image data mining during this period of intensive research and experiments, highlighted by such breakthrough technologies as wavelets, MPEG, Google image search, convolutional neural network, machine learning, ImageNet, Matlab toolboxes, etc.

Given the complexity of image data mining, there is a need for a deep analysis of and insight into the field, especially the latest development, to help researchers understand opportunities and challenges in the field. This book timely captures and presents cutting-edge techniques in the field of image data mining as well as foundational know-how for understanding them. This book provides a complete recipe for image data mining and is a treasure of techniques on image analysis/understanding, feature extraction, machine learning, and image retrieval. The book is built upon the author's career-long and high-impact research in the frontier of this exciting research field. Theories and concepts in the book are typically formulated with practical mathematical models which are realized using algorithms, real data from actual experiments, or working examples. Students and researchers in mathematics and the broader science disciplines will be able to use this book to understand the actual problems/applications in this field and gain firsthand experience of computing. Students and researchers in many areas of the computing discipline will be able to use this book to understand how fundamental and advanced maths are applied to solve the variety of computing problems.

Churchill, Australia

Dengsheng Zhang

About This Book

The book covers the complete know-how on image data mining including math tools, analysis, features, learning, and presentation. It has been organized into four parts: fundamentals, feature extraction, image classification, and image retrieval.

Part I of the book aims to equip readers with some essential tools for image mining. Specifically, Part I provides a brief and evolutional journey from the classical Fourier transform (Chap. 1) to Gabor filters (Chap. 2) and to contemporary wavelet transform (Chap. 3). It prepares readers with fundamental math for some of the advanced mining techniques discussed in the book. Apart from the theories, this part also uses Fourier spectra, STFT spectrogram, Gabor filter spectra, and wavelet spectra to demonstrate how key information or features in an image can be captured by these fundamental transforms.

Parts II and III are the core of the book, which examine and analyze varieties of state-of-the-art models, tools, algorithms/procedures, and machines for image mining. In contrast to Part I which is mostly theoretical, these two parts focus on dealing with real image data and real image mining. Part II demonstrates how a variety of features can be mined or extracted from images for image representation; it covers three chapters which focus on color (Chap. 4), texture (Chap. 5) and shape (Chap. 6), respectively. Each chapter typically begins with simple methods or methods at the intro level and moves on to the more advanced methods in a natural flow. Most of the methods in Part II are demonstrated with intuitive illustrations.

If Part II is analog to raw mining, Part III is about refinery. Specifically, Part III presents readers with four powerful learning machines to classify image data, including Bayesian (Chap. 7), SVM (Chap. 8), ANN/CNN (Chap. 9) and DT (Chap. 10). Each chapter in this part begins with an icebreaking and introductory journey to give readers a big picture and an orientation to follow. It then navigates to the more advanced topics with illustrations to key concepts and components of the learning machine. The story of each machine learning method is typically told with concise maths, demonstrations, applications, and implementations.

After a breathtaking and arduous journey on image mining involving feature extraction and machine learning, readers are soothed with a recovering journey on image retrieval in Part IV. Part IV deals with putting images in order, inspecting the quality of a haul and organizing them for presentation or display. Indexing techniques suitable for image data are first described in detail in Chap. 11 followed by the analysis of a number of image ranking techniques in Chap. 12. The part concludes in Chap. 13 with a number of interesting image presentation techniques and powerful image database visualization methods.

Key Features of the Book

A shortcut to AI. AI and machine learning are usually intimidating to many who don't have the sophisticated mathematics background. This book, however, offers readers a surprising shortcut to AI on machine learning by introducing four major machine learning tools with filtered and easy to understand mathematics using rich illustrations.

A natural marriage between maths and data. Maths and data can only be understood well when they are well matched. This book brings mathematics and computing into a single display and tells image stories with maths by a trained mathematician.

Visualization of image data mining. With more than 200 illustrations (multiple illustrations in some figures), it can be said that the book is a visualization of image data mining, making it very easy to read and understand for readers.

End of chapter summary. Every chapter of the book is equipped with an end of chapter summary to highlight the key points and connect the dots in the chapter.

Exercises. High-quality exercises with instructions or Matlab code have been created for most of the chapters in the book, giving readers the opportunities to test their skills learnt from the book.

Writing for scanning. The book makes extensive use of powerful techniques for scientific and academic writing including inverted pyramid writing, bullet lists, plain language, keyword headings, text chunking, analogy, scannable content, blurbs, etc. Due to writing for scanning, it makes reading the book very efficient and a good experience.

Contents

About the Author

Dr. Dengsheng Zhang received his Ph.D. in computing from Monash University, Australia, in 2002. He has also earned a Bachelor of Science in Mathematics in 1985 and a Bachelor of Arts in English in 1987. He is currently a Senior Lecturer at School of Science, Engineering and Information Technology in Federation University Australia. Dr. Zhang is also a guest professor of Xi'an University of Posts & Telecommunications.

Dr. Zhang has over 25 years of research experience in the field of artificial intelligence and big data analysis. His main research interests include pattern recognition, machine learning, multimedia data classification, and retrieval. He has published more than 100 refereed papers on international journals and conference proceedings in his career. His publications have so far attracted nearly 10,000 citations according to Google Scholar. One of his papers published in Pattern Recognition in 2007 has been awarded the Best Journal Article by the International Association of Pattern Recognition (IAPR) in CVPR2010.

Dr. Zhang has won two Australian Research Council (ARC) Discovery Grants to support his research on image and music data classification. He is also an experienced Ph.D. supervisor.

Dr. Zhang has over 30 years of international teaching experience and has lectured extensively on both undergraduate and postgraduate courses including mathematical analysis, business statistics, business programming, web design and development, multimedia development, computer systems, computer networks, e-commerce, computer interface design and development, mobile agent, business systems, general operations, professional development, computer models for business decisions and project management, etc.

List of Figures

List of Tables

A good beginning is half done.

In image analysis and understanding, no other tools are more important than wavelet transform. Wavelet is a remarkable achievement of decades of research on signal processing and analysis. Although many books have been written on wavelet, they are often too focused on the theoretical part and too mathematical. This has hindered many researchers especially novice researchers from good understanding and application of this essential and important tool.

Wavelet theory is complex, and it is difficult to understand it without rigorous mathematical training. Since wavelet theory is based on Fourier series and Fourier transform, it is impossible to understand how wavelet works without understanding how Fourier transform works. On the other hand, a good understanding of Fourier transform will naturally lead to the understanding of wavelet theory.

In this part of the book, we present the wavelet transform theory as an evolution from Fourier transform. We start with Fourier series and Fourier transform by preparing readers with mathematical fundamentals and building a foundation for understanding wavelet. More importantly, the connection between Fourier theory and its application on signal processing is systematically shown to the readers. By finishing the Fourier transform chapter, readers will be able to understand the limitations of Fourier transform, and this naturally leads to Short-Time Fourier Transform (STFT) and Gabor transform.

Once completing both Chaps. 1 and 2, readers are fully prepared to understand wavelet transform. Wavelet transform is just a natural extension of STFT and Gabor transform. To further strengthen readers understanding, the implementation of wavelet transform is shown by both theoretical demonstrations and practical examples.

This part is the foundation of the many techniques discussed in this book. By completing this part, readers will be able to understand those powerful feature extraction methods including the curvelet transform in Part II.

Fourier Transform

1

In essence, the world is just elements and compounds.

1.1 Introduction

Fourier transform has played a key role in image processing for many years, and it continues to be a topic of interest in theory as well as application. The fundamental principle behind Fourier transform is that a pattern can be treated as a signal, and as such, it can be represented by elementary components of the signal. If we can define elementary components to represent or approximate a pattern under analysis, we can determine how significant an elementary component in a given pattern. The elementary components found in the signal can be used to describe the given pattern. Fourier transform is useful for pattern analysis and description because different patterns can be distinguished by the transformed spectra (Fig. 1.1) [1], while similar patterns will have similar transformed spectra even they are affected by noise and other variations. It can be observed that the spectrum of Fig. 1.1a clearly shows patterns of both horizontal and vertical directions, while that in Fig. 1.1b shows patterns of random fashion. They demonstrate the power of Fourier transform in image analysis and understanding.

1.2 Fourier Series

1.2.1 Sinusoids

To understand how Fourier transform works, it has to start with understanding how sinusoids work. It is important to understand the relationship between frequency and period. Figure 1.2 shows a sine wave and its harmonic waves. It shows how the change of variable scaling or horizontal stretching affects the sine wave's frequency and periods.

© Springer Nature Switzerland AG 2019
D. Zhang, *Fundamentals of Image Data Mining*, Texts in Computer Science, https://doi.org/10.1007/978-3-030-17989-2_1

(a)   **(b)**

Fig. 1.1 Fourier spectra of different images. **a** A scenic image at the left and its Fourier spectrum at the right; **b** a tree image in the left and its Fourier spectra in the right. The brighter the pixel, the higher magnitude of the spectrum

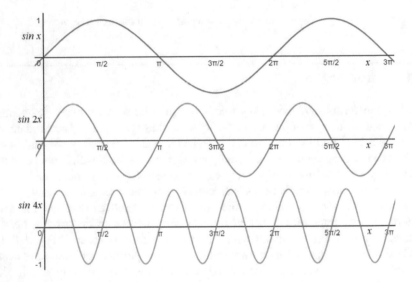

Fig. 1.2 Three sine waves $\sin(nx)$ with different periods and frequencies

Sine waves	Frequency	Periods
$\sin x$	$\frac{1}{2\pi}$	2π
$\sin 2x$	$\frac{1}{\pi}$	π
$\sin 4x$	$\frac{2}{\pi}$	$\frac{\pi}{2}$
$\ldots$		
$\sin nx$	$\frac{n}{2\pi}$	$\frac{2\pi}{n}$

As can be seen, as the variable scaling factor n increases, the period of $\sin(nx)$ becomes shorter and the frequency becomes higher. For example, the period of sin (x) is 2π, while the periods of $\sin(2x)$ and $\sin(4x)$ are π and $1/2\pi$, respectively. Consequently, the frequencies of the three sine waves are $1/2\pi$, $1/\pi$, and $2/\pi$, respectively. Similarly, the period and frequency of $\sin(nx)$ are $2\pi/n$ and $n/2\pi$, respectively.

The sine waves expressed this way do not have an easy interpretation of the frequency and periods, because both of them are in angular terms. Now let's replace n with $2\pi n$ and change the sine function from $\sin(nx)$ to $\sin(2\pi nx)$, see what will happen.

Sine waves	Frequency	Periods
$\sin(2\pi x)$	1	1
$\sin(4\pi x)$	2	$\frac{1}{2}$
$\sin(8\pi x)$	4	$\frac{1}{4}$
$\cdots$		
$\sin(2\pi nx)$	n	$\frac{1}{n}$

Now both the periods and frequencies are easier to understand. For example, the period of $\sin(2\pi x)$ is 1, while the periods of $\sin(4\pi x)$ and $\sin(8\pi x)$ are $1/2$ and $1/4$, respectively. Consequently, the frequencies of the three sine waves are 1, 2, and 4, respectively (Fig. 1.3). Therefore, the period and frequency of $\sin(2\pi nx)$ are $1/n$ and n, respectively. This is much easier to interpret.

A more general form of sine function is expressed as $\sin(2\pi n/L)$, which has a period of L/n and frequency of n/L. This is extremely helpful to analyze signals with arbitrary periodicity and frequencies.

Sine waves	Frequency	Periods
$\sin\left(\frac{2\pi x}{L}\right)$	$\frac{1}{L}$	L
$\sin\left(\frac{4\pi x}{L}\right)$	$\frac{2}{L}$	$\frac{L}{2}$
$\sin\left(\frac{8\pi x}{L}\right)$	$\frac{4}{L}$	$\frac{L}{4}$
$\cdots$		
$\sin\left(\frac{2\pi nx}{L}\right)$	$\frac{n}{L}$	$\frac{L}{n}$

1.2.2 Fourier Series

One of the most important and interesting discoveries in mathematics is that any math function can be approximated with a *series of sinusoids* (sine and cosine waves), called *Fourier series*. Now consider a signal function $f(x)$ with period L,

$$f(x) = a_0 + \sum_{n=1}^{\infty} \left(a_n \cos\frac{2\pi nx}{L} + b_n \sin\frac{2\pi nx}{L}\right) \tag{1.1}$$

To determine the Fourier coefficients a_n and b_n, we multiply both sides of the above equation with either $\sin\left(\frac{2\pi nx}{L}\right)$ or $\cos\left(\frac{2\pi nx}{L}\right)$ and do the integral in $[-L/2, L/2]$.

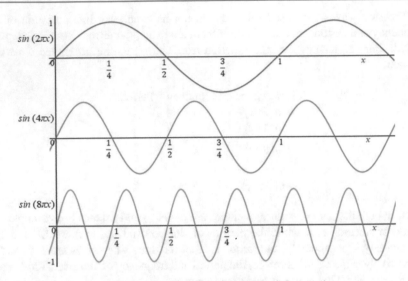

Fig. 1.3 Three sine waves sin($2\pi nx$) with different periods and frequencies

It can be shown that sine and cosine waves have the following convenient properties:

$$\int\limits_{-L/2}^{L/2} \cos\frac{2\pi nx}{L}\cos\frac{2\pi mx}{L}\,dx = \begin{cases} L/2 & for\, n=m \\ 0 & for\, n\neq m \end{cases} \qquad (1.2)$$

$$\int\limits_{-L/2}^{L/2} \sin\frac{2\pi nx}{L}\sin\frac{2\pi mx}{L}\,dx = \begin{cases} L/2 & for\, n=m \\ 0 & for\, n\neq m \end{cases} \qquad (1.3)$$

$$\int\limits_{-L/2}^{L/2} \sin\frac{2\pi nx}{L}\cos\frac{2\pi mx}{L}\,dx = 0 \qquad (1.4)$$

$$\int\limits_{-L/2}^{L/2} \sin\frac{2\pi nx}{L}\,dx = 0 \qquad (1.5)$$

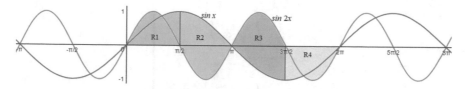

Fig. 1.4 Illustration of $\int \sin x \sin(2x)dx = 0$

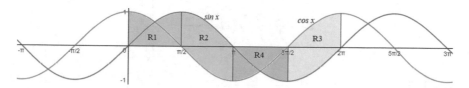

Fig. 1.5 Illustration of $\int \sin x \cos x dx = 0$

$$\int_{-L/2}^{L/2} \cos\frac{2\pi nx}{L}\, dx = 0 \qquad (1.6)$$

To prove the above properties, we only need to demonstrate that $\int \sin(nx)\sin(mx)dx = 0$ $(m \neq n)$ and $\int \sin(nx)\cos(mx)dx = 0$; the others are obvious due to the symmetry of sine and cosine waves. Without loss of generality, we just need to show $\int \sin x \sin(2x)dx = 0$ and $\int \sin x \cos x\, dx = 0$.

Figure 1.4 illustrates $\int \sin x \sin(2x)dx$ in one period. We divide the *Sum of Product* (SoP) of the two functions within one period into four regions: R1–R4, marked with different colors. It is easy to observe that the SoP magnitude of the four regions is exactly the same; however, the signs of the four corresponding SoPs are +, −, −, and +, respectively, resulting in the total SoP of the period as 0. Applying this to all the other periods, it can be shown $\int \sin x \sin(2x)dx = 0$ on the entire x axis.

Figure 1.5 illustrates $\int \sin x \cos x\, dx$ in one period. Similar to the above, we divide the Sum of Product (SoP) of the two functions in the single period into four regions: R1–R4, marked with different colors. Again, it is easy to observe that the SoP magnitude of the four regions are exactly the same, however, the signs of the 4 corresponding SoPs are +, −, + and −, respectively, resulting in the total SoP of the period as 0. Applying this to all other periods, it can be shown $\int \sin x \cos x\, dx = 0$ on the entire x axis.

By making use of the integral identities and orthogonality of (1.2)–(1.6), the Fourier coefficients are obtained as follows:

$$a_0 = \frac{1}{L} \int\limits_{-L/2}^{L/2} f(x)dx$$

$$a_n = \frac{2}{L} \int\limits_{-L/2}^{L/2} f(x)\cos\frac{2\pi n x}{L}dx \Bigg\}$$ (1.7)

$$b_n = \frac{2}{L} \int\limits_{-L/2}^{L/2} f(x)\sin\frac{2\pi n x}{L}dx$$

$n = 1, 2, \ldots$

1.2.3   Complex Fourier Series

Using the Euler formula

$$e^{ix} = \cos x + i\sin x$$ (1.8)

where $i = \sqrt{-1}$. It is easy to work out

$$\cos(x) = \frac{1}{2}\left(e^{ix} + e^{-ix}\right)$$ (1.9)

$$\sin(x) = \frac{1}{2i}\left(e^{ix} - e^{-ix}\right) = -\frac{i}{2}\left(e^{ix} - e^{-ix}\right)$$ (1.10)

Now, by replacing the sinusoids in the Fourier series of (1.1) with the above two equations, we obtain the complex Fourier series:

$$f(x) = \frac{a_0}{2} + \sum_{n=1}^{\infty}\left(a_n\cos\frac{2\pi n x}{L} + b_n\sin\frac{2\pi n x}{L}\right)$$

$$= \frac{a_0}{2} + \frac{1}{2}\sum_{n=1}^{\infty}a_n(e^{i\frac{2\pi n x}{L}} + e^{-i\frac{2\pi n x}{L}}) - \frac{i}{2}\sum_{n=1}^{\infty}b_n(e^{i\frac{2\pi n x}{L}} - e^{-i\frac{2\pi n x}{L}})$$ (1.11)

$$= \frac{a_0}{2} + \frac{1}{2}\sum_{n=1}^{\infty}(a_n - ib_n)e^{i\frac{2\pi n x}{L}} + \frac{1}{2}\sum_{n=1}^{\infty}(a_n + ib_n)e^{-i\frac{2\pi n x}{L}}$$

which can be written as the complex Fourier series

$$f(x) = \sum_{n=-\infty}^{\infty} c_n e^{i2\pi nx/L} \tag{1.12}$$

The exponential form of orthogonality is as follows:

$$\int_{-\frac{L}{2}}^{\frac{L}{2}} e^{-\frac{i2\pi mx}{L}} e^{\frac{i2\pi nx}{L}} = \begin{cases} L & for\, m = n \\ 0 & otherwise \end{cases} \tag{1.13}$$

Now by multiplying both sides of the Fourier series (1.12) with $e^{-i2\pi nx/L}$ and do integral in $[0, L]$, we obtain complex Fourier coefficients:

$$c_n = \frac{1}{L} \int_{-\frac{L}{2}}^{\frac{L}{2}} f(x) e^{-i\frac{2\pi nx}{L}} dx, n = 0, \pm 1, \pm 2, \ldots \tag{1.14}$$

1.3 Fourier Transform

Equation (1.14) indicates that the coefficients of the Fourier series are determined by $f(x)$, while (1.12) indicates that $f(x)$ can be reconstructed from Fourier coefficients c_n. Therefore, the Fourier series establish a unique correspondence between $f(x)$ and its Fourier coefficients. Now, consider the integral of (1.14):

$$Lc_n = \int_{-\frac{L}{2}}^{\frac{L}{2}} f(x) e^{-j\frac{2\pi nx}{L}} dx \tag{1.15}$$

where $j = \sqrt{-1}$. If we let $L \rightarrow \infty$, n/L becomes continuous and $n/L \rightarrow u$, (1.15) becomes

$$F(u) = \int_{-\infty}^{\infty} f(x) \exp(-j2\pi ux) \, dx \tag{1.16}$$

Now, by substituting (1.6) into (1.12) and replacing the sum with an integral by using $n/L \rightarrow u$ and $1/L \rightarrow du$, (1.12) becomes

$$f(x) = \int_{-\infty}^{\infty} F(u) \exp(j2\pi ux) \, du \tag{1.17}$$

The $F(u)$ of (1.16) is called the forward *Fourier transform* or FT, and (1.17) is called the *inverse Fourier transform* or FT^{-1}.

1.4   Discrete Fourier Transform

1.4.1   DFT

Discrete Fourier Transform (DFT) is particularly useful for digital pattern analysis, because digital patterns exist in discrete form. To define DFT from Fourier series, f (x) is first discretized into N samples in $[0, L]$:

$$f(0), f(\Delta x), f(2\Delta x), \ldots, f((N-1)\Delta x) \tag{1.18}$$

where Δx is the sample step in spatial domain and $L = N\Delta x$, and then $f(x)$ can be expressed as

$$f(k) = f(k\Delta x), k = 0, 1, 2, \ldots, N-1 \tag{1.19}$$

Now consider the Fourier coefficients (1.14):

$$
\begin{aligned}
c_n &= \frac{1}{L} \int_{-L/2}^{L/2} f(x) e^{-j\frac{2\pi nx}{L}} dx \\
&= \frac{1}{L} \int_{0}^{L} f(x) e^{-j\frac{2\pi nx}{L}} dx
\end{aligned}
\tag{1.20}
$$

By substituting $L = N\Delta x$, $f(x) = f(k)$, $x = k\Delta x$, and $dx = \Delta x$ into the above equation, it yields

$$
\begin{aligned}
c_n &= \frac{\Delta x}{N\Delta x} \sum_{k=0}^{N-1} f(k) e^{-j\frac{2\pi nk\Delta x}{N\Delta x}} \\
&= \frac{1}{N} \sum_{k=0}^{N-1} f(k) e^{-j\frac{2\pi nk}{N}} n = 0, 1, 2, \ldots, N-1
\end{aligned}
\tag{1.21}
$$

Therefore, the DFT of $f(x)$ is given as

$$F(u) = \frac{1}{N} \sum_{x=0}^{N-1} f(x) \exp(-j2\pi ux/N) \quad u = 0, 1, 2, \ldots, N-1 \tag{1.22}$$

By substituting (1.22) into (1.12), the inverse DFT is obtained as

$$f(x) = \sum_{u=0}^{N-1} F(u)\exp(j2\pi ux/N) \quad x = 0,1,2,\ldots,N-1 \qquad (1.23)$$

1.4.2   Uncertainty Principle

Assume $f(x)$ is a signal in a time period of $\Delta T = [0, L]$, the sampling step Δu in frequency domain and the sampling step Δx in spatial domain are related by the following expression:

$$\Delta u = \frac{1}{\Delta T} = \frac{1}{N\Delta x} \qquad (1.24)$$

Basically, (1.24) tells that the frequency sampling step is inversely proportional to the spatial sampling step. This is known as the *uncertainty principle*, which means that increasing spatial resolution (reduce Δx) reduces the frequency resolution and vice versa. In other words, higher spatial resolution and higher frequency resolution cannot be achieved simultaneously. This is the key reason behind the multiresolution analysis such as wavelets which will be discussed later on in Chap. 3.

Since the Δx depends on the sampling rate f_s, and the relationship between Δx and f_s is given by $\Delta x = 1/f_s$, the above inequality becomes

$$\Delta u \geq \frac{f_s}{N} \qquad (1.25)$$

and the uth frequency is given by

$$f_u = u \cdot \frac{f_s}{N} \qquad (1.26)$$

It should be noted that the uth frequency computed from (1.22) is not the actual frequency; instead, the uth frequency is the uth *bin* of frequency. In other words, u is the bin number, and the actual frequency is given by (1.26): $f_u = u \cdot \Delta u$, and Δu is called the bin size of DFT. If $f_s = N$, $\Delta u = 1$, this is often the assumption in DFT. However, this is not always the case. When $f_s \gg N$, $\Delta u \gg 1$, this will be demonstrated in Sect. 2.2. Equation (1.25) is another form of the *uncertainty principle*. Given a sampling rate, in order to increase the frequency resolution (reduce Δu), it has to increase the sample or window size N, which reduces the spatial resolution. This is called the *trade-off* between spatial resolution and frequency resolution.

It should also be noted that any window size of a DFT is relative according to (1.25). Specifically, a window size is relative to the sampling rate or sampling frequency. A window of N samples is smaller in a signal with faster sampling rate than that in a signal with slower sampling rate. For example, in a signal with 44,000 Hz sampling rate, a window of 128 samples has a duration of $128/44,000 = 0.0029$ s. However, in a signal with 22,000 Hz sampling rate, the duration of a 128 samples window is $128/22,000 = 0.0058$ s, which is twice the size as that in the first signal. This indicates that a bin number (u) of a DFT computed from windows of different sizes or different signals means different frequencies.

The inverse relationship between frequency resolution and spatial resolution (window size) can be demonstrated using the following example. Suppose there are two sine waves with very small frequency difference [2]:

$$\text{Sine wave one:} \quad \sin(2\pi \times 0.05x)$$
$$\text{Sine wave one:} \quad \sin(2\pi \times 0.0501x)$$

In this case, $\Delta u = 0.0001$. If we plot the two sine waves in one graph (Fig. 1.6), one in red and the other in blue, the two signals do not show a difference in the first 100 samples, which means a small window cannot discern the difference of the two signals. However, if we show the two signals in a very large window (5,000 samples), at the end of the window, they are 180° out of phase. This is because the periods of the two sine waves are 20 and 19.96, respectively. Assume a sampling frequency of 1 Hz. For a 100 samples window, the difference between the two signals is just $5.01 - 5 = 0.01$ period, which is almost indiscernible. However, with a 5,000 samples window, the difference between the two signals is $250.5 - 250 = 0.5$ period, which is more than sufficient to distinguish the two signals. It is more convenient to explain this case using (1.25), because the smallest frequency difference can be detected in a 100 samples window is $\Delta u = 1/100 = 0.01$, while in

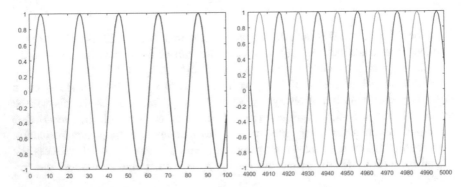

Fig. 1.6 Inverse relationship between spatial and frequency resolution. Left: the first 100 samples of the two sine waves; right: the last 100 samples of the two sine waves

a 5,000 samples window, $\Delta u = 1/5,000 = 0.0002$, which is able to distinguish the two signals.

It demonstrates that a smaller window gives poor frequency resolution, while a larger window gives higher frequency resolution. This is because the larger the window, the more samples, and the more low frequencies can be computed.

1.4.3 Nyquist Theorem

Because frequency is measured by the number of cycles in a period of time, and the smallest cycle consists of two samples, for a signal of size N, only $N/2$ frequencies can be computed from the DFT. This is called the *Nyquist theorem*.

Another way to express the *Nyquist theorem* is that in order to reconstruct/recover a signal appropriately ("appropriately" means recover the "essence" or low frequency while ignoring the "nuance" or high frequency), the sampling rate of the signal must be at least twice the highest frequency in the signal. Figure 1.7 demonstrates this fact. The figure shows three signals of 1 s length. The top signal is a sine wave with 1 cycle/period (frequency = 1) which can be recovered or reconstructed appropriately by at least two samples (marked with red dots). The middle signal is a sine wave with two cycles/periods in a second (frequency = 2); it needs at least four samples to recover the signal appropriately. The

Fig. 1.7 Illustrations of different sampling rates for three signals of the same time length

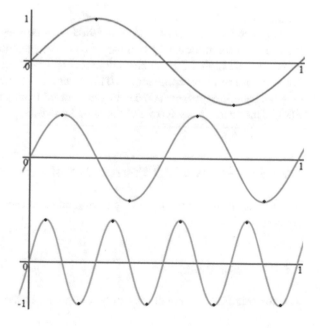

bottom signal is a sine wave with 4 cycles/periods in a second (frequency = 4); it needs at least eight samples to recover the signal appropriately, and so on so forth.

In the case of images, frequency is related to structure size, and small structures are known to have high frequency. Because the smallest structure in an image requires 2 pixels to discern, the highest frequency which can be captured in an image is 1/2 pixels.

1.5 2D Fourier Transform

For a two-variable function $f(x, y)$ defined in $0 \leq x, y < N$, its Fourier transform pair is given by

$$F(u,v) = \frac{1}{N}\sum_{x=0}^{N-1}\sum_{y=0}^{N-1} f(x,y) \exp[-j2\pi(ux+vy)/N] \qquad (1.27)$$

for $u, v = 0, 1, 2, \ldots, N - 1$, and $j = \sqrt{-1}$.

$$f(x,y) = \frac{1}{N}\sum_{u=0}^{N-1}\sum_{v=0}^{N-1} F(u,v) \exp[j2\pi(ux+vy)/N] \qquad (1.28)$$

for $x, y = 0, 1, 2, \ldots, N - 1$.

Although the number of $F(u, v)$ resulted from Fourier transform is usually large, the number of significant $F(u, v)$ (or $F(u)$) (large magnitude) is usually small. This is because the higher frequencies only represent the finest pattern details which are not so useful in many applications. This means that a meaningful approximation of original pattern $f(x, y)$ (or $f(x)$) can be constructed from a small number of $F(u, v)$ (or $F(u)$). This forms the basis of Fourier signal processing and Fourier pattern analysis.

1.6 Properties of 2D Fourier Transform

Fourier transform has the following important properties which are useful for image analysis.

1.6.1 Separability

The discrete Fourier transform can be expressed in the separable form

$$F(u, v) = \frac{1}{\sqrt{N}} \sum_{x=0}^{N-1} \left[\frac{1}{\sqrt{N}} \sum_{y=0}^{N-1} f(x, y) \exp\left(-\frac{j2\pi vy}{N}\right) \right] \exp\left(-\frac{j2\pi ux}{N}\right)$$

$$= \frac{1}{\sqrt{N}} \sum_{x=0}^{N-1} F(x, v) \exp\left(-\frac{j2\pi ux}{N}\right) \tag{1.29}$$

$$= FT_x\{FT_y[f(x, y)]\}$$

where FT_x and FT_y are the 1D FTs on row and column, respectively.

The advantage of the separability is that $F(u, v)$ can be obtained in two steps by successive applications of 1D FT which can be computed using the *Fast Fourier Transform* (FFT).

1.6.2 Translation

The translation property of the Fourier transform is given by

$$FT[f(x - x_0, y - y_0)] = F(u, v) \cdot \exp[-j2\pi(ux_0 + vy_0)/N] \tag{1.30}$$

It indicates that a shift in spatial domain results in a phase change in frequency domain. That means the magnitude of Fourier transform is invariant to translation. This is a desirable feature, because, in many applications, the phase information is discarded which leaves the FT features invariant to translation.

1.6.3 Rotation

To find the relationship between a rotated function $f(x, y)$ and its spectrum, let's assume the function $f(x, y)$ is rotated by an angle θ, and the function after the rotation is $f(x', y')$. Then the relationship between two corresponding points of the two functions is as follows:

$$x' = x\cos\theta + y\sin\theta \tag{1.31}$$

$$y' = y\cos\theta - x\sin\theta \tag{1.32}$$

$$x = x'\cos\theta - y'\sin\theta \tag{1.33}$$

$$y = x'\sin\theta + y'\cos\theta \tag{1.34}$$

By substituting (1.33) and (1.34) into (1.27), we have

$$
\begin{aligned}
F(u',v') &= \frac{1}{N}\sum_{x'=0}^{N-1}\sum_{y'=0}^{N-1} f(x',y')\exp\left[-j2\pi\left(\frac{ux'\cos\theta - uy'\sin\theta + vx'\sin\theta + vy'\cos\theta}{N}\right)\right] \\
&= \frac{1}{N}\sum_{x'=0}^{N-1}\sum_{y'=0}^{N-1} f(x',y')\exp\left[-j2\pi x'\left(\frac{u\cos\theta + v\sin\theta}{N}\right)\right]\exp\left[-j2\pi y'\left(\frac{v\cos\theta - u\sin\theta}{N}\right)\right] \\
&= \frac{1}{N}\sum_{x'=0}^{N-1}\sum_{y'=0}^{N-1} f(x',y')\exp\left[-j2\pi\left(\frac{x'u' + y'v'}{N}\right)\right]
\end{aligned}
$$

$$(1.35)$$

where

$$u' = u\cos\theta + v\sin\theta \tag{1.36}$$

$$v' = v\cos\theta - u\sin\theta \tag{1.37}$$

Therefore, rotating $f(x, y)$ by an angle of θ in spatial domain rotates $F(u, v)$ by the same angle in frequency domain.

The rotation property can be proved more conveniently by considering $f(x, y)$ and FT in either complex domain or polar space. A point (x, y) in complex domain can be expressed as

$$z = x + jy \tag{1.38}$$

By using Euler's formula, it is simple to shown that

$$
\begin{aligned}
ze^{-j\theta} &= (x+jy)\cdot(\cos\theta - j\sin\theta) \\
&= x\cos\theta + jy\cos\theta - jx\sin\theta + y\sin\theta \\
&= (x\cos\theta + y\sin\theta) + j(y\cos\theta - x\sin\theta) \\
&= x' + jy'
\end{aligned}
\tag{1.39}
$$

Equation (1.39) shows that a point z rotated by an angle θ is equivalent to z times $e^{-j\theta}$. In other words, the following is true:

$$f(x',y') = f(x,y)\cdot e^{-j\theta} \tag{1.40}$$

Equation (1.40) is a more concise and convenient rotation formula than (1.31) and (1.32). By substituting (1.40) into (1.27), we obtain the FT of the rotated function $f(x', y')$:

$$F(u', v') = \frac{1}{N} \sum_{x=0}^{N-1} \sum_{y=0}^{N-1} f(x', y') \exp\left[-j2\pi\left(\frac{ux + vy}{N}\right)\right]$$

$$= \frac{1}{N} \sum_{x=0}^{N-1} \sum_{y=0}^{N-1} f(x, y) e^{-j\theta} \exp\left[-j2\pi\left(\frac{ux + vy}{N}\right)\right] \quad (1.41)$$

$$= F(u, v) \cdot e^{-j\theta}$$

Therefore, we obtain the same result as shown in (1.35).

If we consider both $f(x, y)$ and $F(u, v)$ in polar space, they can be expressed as $f(r, \theta)$ and $F(\rho, \phi)$, respectively, where

$$x = r\cos\theta, y = r\sin\theta; \quad u = \rho\cos\phi, v = \rho\sin\phi \quad (1.42)$$

(r, θ) is the polar coordinates in image plane and (ρ, ϕ) is the polar coordinates in frequency plane. The differentials of x and y are

$$\left. \begin{array}{l} dx = \cos\theta \, dr - r\sin\theta \, d\theta \\ dy = \sin\theta \, dr - r\cos\theta \, d\theta \end{array} \right\} \quad (1.43)$$

The Jacobian of (1.43) is r. Therefore, by substituting both (1.42) and (1.43) into 2D continuous FT, the 2D FT in polar space is given by the following equations:

$$F(\rho, \phi) - \int_0^\infty \int_0^{2\pi} f(r, \theta) e^{-j2\pi(r\cos\theta\rho\cos\phi + r\sin\theta\rho\sin\phi)} r \, dr \, d\theta$$

$$= \int_0^\infty \int_0^{2\pi} f(r, \theta) e^{-j2\pi r\rho\cos(\theta - \phi)} r \, dr \, d\theta \quad (1.44)$$

Suppose $f(r, \theta)$ is rotated for an angle of θ_0 to $f(r, \theta + \theta_0)$. Let $\theta' = \theta + \theta_0$, then

$$\theta = \theta' - \theta_0 \quad \text{and} \quad dq = d\theta' \quad (1.45)$$

Now, in (1.44), by substituting $f(r, \theta)$ with $f(r, \theta')$ and substituting θ with (1.45), we obtain

$$F(\rho, \phi') = \int_0^\infty \int_0^{2\pi} f(r, \theta') e^{-j2\pi r\rho\cos[\theta' - (\phi + \theta_0)]} r \, dr \, d\theta' \quad (1.46)$$

Equation (1.46) means

$$FT[f(r, \theta + \theta_0)] = F(r, \phi + \theta_0) \tag{1.47}$$

Again, this yields the same result as (1.35) and (1.41). Equation (1.47) also tells that in polar domain, the rotation of an image causes a translation or shift on its FT spectrum. This property is useful for feature normalization.

1.6.4 Scaling

For two scalars a and b, the scale property of Fourier transform is given by

$$FT[f(ax, by)] = \frac{1}{ab} F\left(\frac{u}{a}, \frac{v}{b}\right) \tag{1.48}$$

It indicates the scaling of $f(x, y)$ with a and b in x and y directions in spatial domain (time domain in 1D case) causes inverse scaling of magnitude of $F(u, v)$ in frequency domain. That means, if you stretch $f(x, y)$ in spatial domain, you shrink $F(u, v)$ in frequency domain and vice versa. This proves the *uncertainty principle* from another perspective. In general terms, enlarging an object in an image gives rise to lower frequencies in spectral domain while shrinking an object in an image gives rise to higher frequencies in spectral domain. This property is useful in dealing with image scaling.

1.6.5 Convolution Theorem

The *convolution theorem* states that the FT of a convolution between two functions is equal to the product of two FTs. Specifically, given two function f and g, the following are true:

$$FT[f * g] = FT[f] \cdot FT[g] \tag{1.49}$$

$$f * g = FT^{-1}\{FT[f] \cdot FT[g]\} \tag{1.50}$$

where $f * g$ means convolution. Because of the separability property of 2D FT, we only need to prove the 1D case.

$$FT[f * g] = \sum_n \sum_m f(m)g(n-m)e^{-\frac{j2\pi nu}{N}}$$

$$= \sum_m f(m) \sum_n g(n-m)e^{-\frac{j2\pi nu}{N}}$$

$$= \sum_m f(m)FT[g]e^{-\frac{j2\pi mu}{N}} \; (translation\, property) \qquad (1.51)$$

$$= FT[g] \sum_n f(m)e^{-\frac{j2\pi mu}{N}}$$

$$= FT[f] \cdot FT[g]$$

Convolution theorem shows that convolution in spatial domain can be done by an FT (FFT in practice) and a product. This is a useful feature because both FFT and product are much more efficient than spatial convolution.

1.7 Techniques of Computing FT Spectrum

The magnitude image of a Fourier transform is called an FT spectrum. The intensity of an FT spectrum has a very large dynamic range; it is impossible to display this large range in a gray level image. For example, the dynamic range of spectral values of the Lena image [3] is [0, 31, 744], and Fig. 1.8a shows the FT spectrum without scaling. It can be seen that the spectral image reveals little information about the input image. Conventional *thresholding* (Fig. 1.8b) and *linear scaling* (Fig. 1.8c) do not work well for such a large range of values.

The common practice of displaying an FT spectrum is to do a *logarithmic* transformation of the spectral values to bring down the large spectral values to well within the display range of 255 and raise the lower spectral values in the meantime. However, the logarithm transformed spectrum does not have sufficient contrast between lower frequency and higher frequency spectral values as shown in Fig. 1.8d. A more effective way to display FT spectrum is to apply a logarithm transform on the spectral values followed by a linear scaling to map the spectral magnitudes to [0, 255] using (1.52):

$$F'(u,v) = 255 \times \frac{\log(1+|F(u,v)|)}{\log[1+\max(|F(u,v)|)]} \qquad (1.52)$$

Figure 1.8e shows the FT spectrum using (1.52). It can be seen from Fig. 1.8e that there are three directional features in the FT spectrum: horizontal, vertical, and diagonal. The strong horizontal feature is due to the vertical pole on the left-hand

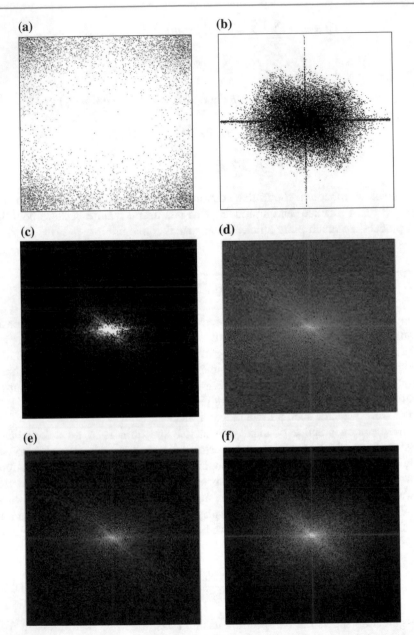

Fig. 1.8 FT spectra with different methods. **a** FT spectrum without scaling; **b** FT spectrum with thresholding value 10; **c** FT spectrum with linear scaling; **d** FT spectrum with log transform; **e** FT spectrum with both log and linear transform; and **f** FT spectrum with enhanced contrast from (**e**)

(a)

(b)

(c)

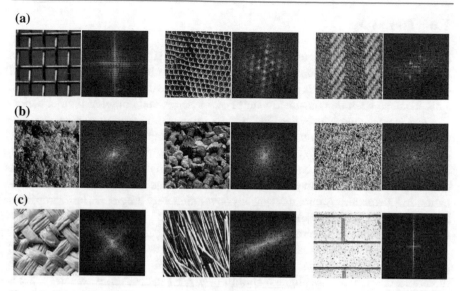

Fig. 1.9 FT spectra of different types of patterns. **a** Regular patterns and their FT spectra on the right; **b** random patterns and their FT spectra on the right; and **c** directional patterns and their FT spectra on the right

side of the input image, while the vertical and diagonal features are due to the rim of the hat and the black arch on the right-hand side of the input image.

The logarithmic transformation, however, enhances the low magnitude values, while compressing high magnitude values into a relatively small pixel range. Therefore, if an image contains some important high magnitude information, this may lead to loss of information. An alternative solution to further increase the spectral contrast is to decrease the compression rate by scaling down the spectrum image intensity before applying the logarithmic transform. This is because the logarithmic function has a less degree of compression at places close to the origin. Figure 1.8f shows the FT spectrum with enhanced contrast, which is equivalent to highlighting the low-frequency area with a spotlight.

The FT spectrum reveals key information about an image if displayed properly. Figure 1.9 shows three different types of homogenous patterns and their FT spectra on the right-hand side of the patterns. It can be seen that the FT spectra have generally accurately captured the three types of texture features: regularity, randomness, and directionality. This is the primary motive for the development of short-time FT and wavelets, which attempt to capture local and changing patterns.

1.8   Summary

In this chapter, FT, the most fundamental theory in signal and image analysis is formally introduced and described systematically. The idea of decomposing a signal or function into sinusoids is equivalent to breaking down compounds into elements. This idea has been demonstrated using Fourier series and complex Fourier series.

Particularly important is the DFT and its properties including rotation, translation, separability, and convolution theorem. For example, due to the separability property, FT on an image can be computed on rows first and then on columns, instead of on rows and columns simultaneously; this makes FT very efficient. Another example is that the convolution theorem turns convolution into multiplication in FT domain. Again, this makes convolution very efficient. These properties have been used frequently in the following five chapters. Another important point is the trade-off between frequency resolution and time/spatial resolution.

FT spectrum is the visualization of FT; it is used to demonstrate what type of patterns are in the image (e.g., random, directional, regular, etc.) and which directions the patterns are aligned. Unfortunately, FT cannot tell where the patterns are located in an image; this is the topic of the next two chapters.

1.9   Exercises

1. Download the stripe image from https://homepages.inf.ed.ac.uk/rbf/HIPR2/ images/stp1.gif, apply FT to it, and show the FT spectrum of this image. The Matlab function for 2D FT is available from https://au.mathworks.com/help/ matlab/ref/fft2.html.
2. Now apply logarithm transform to the above FT spectrum and show the FT spectrum image again. Explain the difference.
3. Apply thresholding to the above log FT spectrum with a threshold value 10 and display the FT spectrum image again. Try different threshold values to see the results.
4. Now multiply the FT spectrum from problem 1 by a circle with a radius of 30 pixels and set the pixels outside the circle with 0 (1 if inside the circle), apply the inverse FT to the truncated spectrum, and show the reconstructed image. Explain the difference between the reconstructed stripe image with the original stripe image. Try circles with a smaller radius, e.g., 10, 5, and explain the results.
5. Repeat Exercises 1 and 4 on different images, write a short report on your findings of FT and FT^{-1}, e.g., why are some images can be reconstructed well with just a few FT coefficients while some images need a high number of FT coefficients to be reconstructed reasonably well.

References

1. Zhang D (2002) Image retrieval based on shape. PhD thesis, Monash University
2. Clayh J, Why does a narrower window in a Fourier Transform or STFT give poor frequency resolution? https://dsp.stackexchange.com/questions/14437/why-does-a-narrower-window-in-a-fourier-transform-or-stft-give-poor-frequency-re/14439. Accessed Feb 2019
3. Wikipedia.org, Lenna. Accessed Feb 2019

Windowed Fourier Transform

2

It's a changing world, static is not an option.

2.1 Introduction

2D Fourier transform is a powerful tool to capture the frequency information of an image. The frequency information tells how frequent a pattern changes. This frequency of changes reflects the structural or textural features which are observed by human beings during pattern analysis. The frequency information is crucial to understand the content of an image.

However, Fourier spectrum is captured using the entire image as the window, it is a global information. In other words, we know there is a frequency in the image, but we cannot tell where the frequency is in the pattern. This is not a problem if the pattern has a homogenous structure across the pattern. For non-homogenous patterns, however, Fourier spectrum is not an effective representation, because different patterns can have similar Fourier spectrum. Figure 2.1 shows this phenomenon [1], although the two images are very different, however, their FT spectra are quite similar. This is a problem for image classification and retrieval. Therefore, we need a better tool to let us have a closer look at the patterns inside the images.

2.2 Short-Time Fourier Transform

The natural way to overcome this problem is to analyze the signal section by section or window by window. This is Short-Time Fourier Transform (STFT) which provides a way to analyze the signal in both time and frequency. In STFT, a window function is chosen in such a way that the portion of a nonstationary signal which is covered by the window function seems stationary. This window function is then convoluted with the original signal so that only the part of the signal covered by the window is selected. FT is then applied to the newly generated stationary

© Springer Nature Switzerland AG 2019
D. Zhang, *Fundamentals of Image Data Mining*, Texts in Computer
Science, https://doi.org/10.1007/978-3-030-17989-2_2

Fig. 2.1 Two images and their corresponding Fourier spectra on the right

signal. The window is then moved to the next slot of signal, and FT is applied repeatedly until the whole signal is completely analyzed. For signal $f(x)$, its STFT is defined as

$$STFT(\tau, \omega) = \sum_{x=0}^{N} f(x) * W(x - \tau)e^{-j2\pi\omega x} \qquad (2.1)$$

where $W(x)$ is the window, $*$ means convolution, τ represents the spatial position of the window, and ω represents the frequency captured at time τ. Similarly, the 2D STFT is given as

$$STFT(\tau_1, \tau_2, \omega) = \sum_{x=0}^{N} \sum_{y=0}^{N} f(x, y) * W(x - \tau_1, y - \tau_2)e^{-j2\pi\omega x} \qquad (2.2)$$

Figure 2.2 shows the different spectrum layouts of FT and STFT on a 1D signal. The FT is applied on the entire signal which is equivalent to a single big window; it can be seen that the frequency resolution is higher than that of STFT. STFT is applied on four smaller windows, as can be expected the frequency resolution is lower than that of FT; however, the spatial resolution is higher than that of FT,

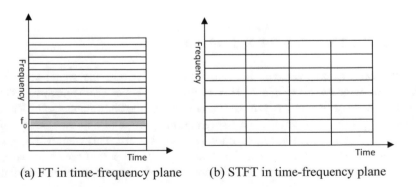

(a) FT in time-frequency plane   (b) STFT in time-frequency plane

Fig. 2.2 Time–frequency illustration for FT and STFT

because we can now examine the signal at four different locations. Therefore, STFT achieves a *trade-off* between frequency resolution and spatial resolution.

2.2.1   Spectrogram

When a signal $f(x)$ ($x \in [0, T]$) is analyzed by STFT, instead of a single FT spectrum, it results in a series of STFT spectra. Each of the STFT spectra is a windowed analysis of the signal $f(x)$ in a particular time slot $t \in [0, T]$. By concatenating the series of the STFT spectra vertically (in column) on the timeline, it creates a *spectrogram*. Figure 2.3 shows a spectrogram of a short sound wave. It can be observed that most of the energy is concentrated at the low frequencies; however, there are a number of particular high frequencies at different times of the sound, which are marked by the bright horizontal stripes.

Although STFT lets us do time–frequency analysis, the usually square windowing causes several side effects. First, the windowing causes the loss of low frequencies which are the most important information for signal representation. This is because low-frequency signals have longer periods/cycles, and in order to capture the low frequency, a signal must complete at least one full cycle within the window. Therefore, a window can only capture frequencies up to a certain limit. For example, given a signal with a Nyquist sampling rate of 44,800 Hz:

- A window of 128 samples is equivalent to a period of 128/44,800 = 0.00285 s.
- Therefore, the lowest frequency the window can capture is 1/0.00285 s = 350 Hz.

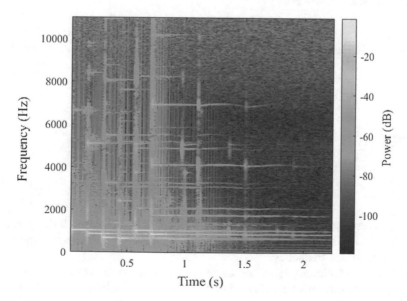

Fig. 2.3 The spectrogram of a sound wave

- In other words, the lowest frequency you can analyze with a window size of 128 samples at a sample rate of 44.8 kHz is 350 Hz.
- The second frequency which can be fit into the window is a two-cycle sine wave or a sine wave with a period of half of the window size; therefore, the second frequency a 128 window captures is $2 \times 350 = 700$ Hz.
- Similarly, the third frequency a 128 window captures is $3 \times 350 = 1,050$ Hz, so on so forth.
- In other words, the step size of the windowed frequency resolution (bin size Δu) is 350 Hz for a 128 window instead of 1 Hz for an ordinary FT. This has been shown in (1.25).
- Similarly, for a window of 64 samples, $\Delta u = 700$ Hz, while for a window of 256 samples, $\Delta u = 175$ Hz, so on so forth.
- Therefore, with STFT, we not only lose low frequencies but also lose frequency resolution, due to using only a single sized window.

Another issue with STFT is the shape of the window. The typical rectangular window causes severe frequency leakage, that is, a burst of high frequencies at both sides of the window. This is undesirable for signal or image representation which requires a compact spectrum. These issues related to STFT can be overcome to a certain extent by using non-rectangular and overlapping windows.

2.3 Gabor Filters

2.3.1 Gabor Transform

This leads to the use of Gaussian window which attenuates high frequencies at both sides of the window. The STFT with Gaussian window is called Gabor transform:

$$G(\tau_1, \tau_2, \omega) = \sum_{x=0}^{N} \sum_{y=0}^{N} f(x,y) * g(x - \tau_1, y - \tau_2) e^{-j2\pi\omega x} \qquad (2.3)$$

where $g(x, y)$ is the Gaussian function:

$$g(x,y) = \frac{1}{2\pi\sigma_x\sigma_y} \exp\left[-\frac{1}{2}\left(\frac{x^2}{\sigma_x^2} + \frac{y^2}{\sigma_y^2}\right)\right] \qquad (2.4)$$

and σ_x, σ_y are the horizontal and vertical standard deviations which determine the size of the window. The window size can be varied to achieve the optimality between time and frequency.

Because convolution in spatial domain is equivalent to multiplication in frequency domain, in practice, STFT is computed by multiplying the Fourier transforms of $f(x, y)$ and $g(x, y)$.

It is found that the frequency response or Fourier transform of $g(x, y)$ is also a Gaussian $G(u, v)$, and the window size of $G(u, v)$ is inversely proportional to that of $g(x, y)$, that is,

$$\sigma_u = \frac{1}{2\pi\sigma_x} \tag{2.5}$$

$$\sigma_v = \frac{1}{2\pi\sigma_y} \tag{2.6}$$

This relationship can be used to determine the window size in spatial domain. It is known that lower frequencies are more important than higher frequencies for signal analysis and representation. Therefore, in the frequency plane, lower frequencies are given higher resolution than higher frequencies. This is achieved by giving the lower frequencies narrower bandwidth while giving the higher frequencies wider bandwidth. Typically, the bandwidths are arranged in *octave*.

2.3.2   Design of Gabor Filters

Because both Gabor function and its frequency response are Gaussians, and the relationship of the two Gaussians is given by (2.5) and (2.6), Gabor filters are designed on frequency domain. Because a 2D Gaussian function extends to infinity, there is too much overlap or redundancy between two adjacent Gaussian functions. To remove the redundancy, the 2D Gaussian functions in Gabor filters are cut at the half height, and the top half of the function is used as the Gaussian window. For a Gaussian function with standard deviation of σ, the Full Width at Half Maximum (FWHM) is $2\sqrt{2\ln 2}\sigma$ (Fig. 2.4).

Fig. 2.4 The full width at half maximum of a Gaussian function

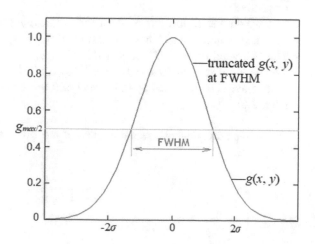

Fig. 2.5 The half-amplitude
of Gabor filters in the
frequency domain using four
scales and six orientations

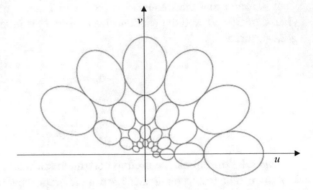

Fig. 2.6 Bandwidth tiling in
frequency plane using
Gaussian windows

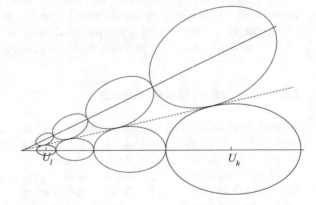

Based on the above discussions, the half-amplitude of Gabor filters tiling of the
spectrum plane is given in Fig. 2.5.

- Suppose the lowest and highest horizontal frequencies are U_l and U_h,
  respectively.
- The window at U_l has the smallest width (bandwidth) σ_l.
- The next window at aU_l has a width of $a\ \sigma_l$.
- The mth window at $a^m U_l$ has a width of $a^m\ \sigma_l$.
- The width of the window at $U_h = a^{M-1}U_l$ is $a^{M-1}\ \sigma_l$.
- The octave is then rotated at an interval of $\theta = \pi/k$ to tile the half frequency plane
  (Fig. 2.6).

With this arrangement, the parameters of the window at are obtained as follows
[2]:

$$\sigma_u = \frac{a-1}{a+1} \cdot \frac{U_l}{\sqrt{2\ln 2}} \tag{2.7}$$

$$\sigma_v = \tan\left(\frac{\pi}{2k}\right) \sqrt{\frac{U_h^2}{2\ln 2} - \sigma_u^2} \qquad (2.8)$$

The Gabor transform lets us do a better time and frequency analysis than STFT, due to the use of Gaussian and overlapping windows. However, because the Gabor function is an infinite window, there is much overlap between Gabor windows. This translates to redundancy in the extracted information from the transformed coefficients. Although the FWHM truncation reduces the redundancy, it causes missing spectral information in frequency plane. Neither case is desirable for image analysis and representation. To overcome this issue, orthogonal wavelets with multiresolution are introduced in the following section.

2.3.3 Spectra of Gabor Filters

Based on the above design, each Gabor filter is determined by two parameters: scale (σ or bandwidth) and orientation; therefore, by changing the scales and orientations,

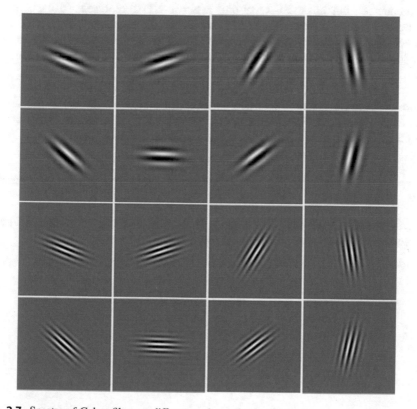

Fig. 2.7 Spectra of Gabor filters at different scales and orientations

various Gabor filters are generated. Figure 2.7 shows the real components of Gabor filters at different scales and orientations. On 2D plane, each Gabor filter is oval-shaped, and the center of the filter is given higher weight. The scale determines the granularity, with lower scale filters capturing rough features of an image and higher scale filters capturing fine features of an image. The orientation let the filters capturing image profiles and edges from different angles. The combination of both scales and orientations provides Gabor filters a powerful capability on image analysis.

2.4 Summary

In this chapter, two windowed FT methods are introduced and discussed in detail. Both STFT and Gabor filters allow for time/space–frequency analysis. Because of using shifting windows, the output of STFT on a 1D signal is a 2D spectrogram instead of a single 1D FT spectrum. It can be observed that the spectrogram reveals a lot more frequency information than a single FT spectrum. However, due to the use of windows, we sacrifice some frequency resolution. That means, instead of 1 frequency per bin in an FT spectrum, a bin in STFT represents a band of frequencies. The bandwidth of an STFT bin depends on the window size; the narrow the window, the wider the bin. We also lose some low frequencies due to windowing. Therefore, it is important to learn the trade-off between window size and bin size when using STFT.

Compared with STFT, Gabor filters provide a better solution in terms of the trade-off, because Gabor filters use multiple filter size. Furthermore, the use of Gaussian window by Gabor filters produces more desirable results than the rectangular window.

2.5 Exercises

1. Match each of the following signals to its corresponding spectrogram underneath the signals and explain why you match it that way.

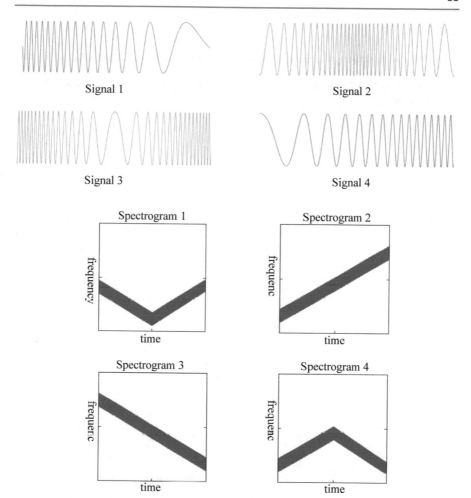

2. Contrast the similarity and difference between STFT and Gabor filters.
3. Explain the different tiling of spectrum plane between STFT and Gabor filters.
4. Use the Matlab code from the following web page to generate Gabor filters: https://au.mathworks.com/help/images/ref/gabor.html. Try different number of scales and orientations and explain how they are different and what kind of image feature they capture.
5. Find a texture image, apply the Gabor filters you have generated in Exercise 4 on the texture image, and explain the spectra of the Gabor filtered images. Try more images and report your findings.

References

1. Sumana I (2008) Image retrieval using discrete curvelet transform. Master thesis, Monash University
2. Rubner Y (1999) Perceptual metrics for image database navigation. PhD thesis, Stanford University

Wavelet Transform

3

To understand the realworld zoom in zoom out.

3.1 Discrete Wavelet Transform

Gabor transform can be written as

$$G_{\sigma_x \sigma_y}(\tau_1, \tau_2) = \sum_{x=0}^{N} \sum_{y=0}^{N} f(x,y) g^*_{\sigma_x \sigma_y}(x - \tau_1, y - \tau_2) \tag{3.1}$$

where * means complex conjugate and the Gabor function $g_{\sigma_x, \sigma_y}(x,y)$ is given as

$$g_{\sigma_x \sigma_y}(x,y) = \frac{1}{2\pi \sigma_x \sigma_y} \exp\left[-\frac{1}{2}\left(\frac{x^2}{\sigma_x^2} + \frac{y^2}{\sigma_y^2} \right) \right] e^{j2\pi \omega x} \tag{3.2}$$

The group of Gabor functions $g_{\sigma_x, \sigma_y}(x,y)$ are windowed waveform functions, called wavelets. But the Gabor wavelets are not orthogonal, which means there is a correlation between different Gabor wavelets. This correlation results in redundancy in the extracted wavelet features computed from images or signals. The FWHM approach in Sect. 2.3 causes loss of frequency. This is undesirable for image or signal representation. The window size is also an issue similar to that of STFT. These issues can be overcome by using orthogonal wavelets with varying window size.

The general form of a 2D orthogonal wavelet can be formulated as follows:

$$\psi_{a_1 a_2 b_1 b_2}(x,y) = \frac{1}{\sqrt{a_1 a_2}} \psi\left(\frac{x - b_1}{a_1}, \frac{y - b_2}{a_2} \right) \tag{3.3}$$

where a_1, a_2 are the scale parameters and b_1, b_2 are the position parameters. Similar to a 2D FT, a 2D wavelet also has the property of separability:

D. Zhang, *Fundamentals of Image Data Mining*, Texts in Computer Science, https://doi.org/10.1007/978-3-030-17989-2_3

$$\psi_{a_1 a_2 b_1 b_2}(x, y) = \frac{1}{\sqrt{a_1}} \psi\left(\frac{x - b_1}{a_1}\right) \frac{1}{\sqrt{a_2}} \psi\left(\frac{y - b_2}{a_2}\right) \tag{3.4}$$

The *Discrete Wavelet Transform* (DWT) on a function or image $f(x, y)$ is given as

$$W^s(k, l) = \frac{1}{s} \sum_n \sum_m f(m, n) \psi\left(\frac{m - k}{s}\right) \psi\left(\frac{n - l}{s}\right) \tag{3.5}$$

where (k, l) is the position of the wavelet and s is the scale. If we compare a wavelet with a magnifying glass, the position vector (k, l) represents the location of the magnifying glass and the scale s represents the distance between the magnifying glass and the image. By adjusting the position and scale, the wavelet can analyze an image in the same way as we analyze an image using a magnifying glass.

3.2   Multiresolution Analysis

As explained in Sect. 1.4, the window size (time) and the frequency band are inversely proportional. That is, when the window size is halved, the frequency band captured by the window is twice higher. The frequency bandwidth is typically arranged in octave, that is, the next bandwidth is twice the width of the previous one. The inverse relationship between window size and frequency of a DWT can be demonstrated using a 1D signal with a Nyquist sampling rate of 1,024 Hz. As can be seen in the following table, as the wavelet decomposition level (scale) and the window size increases, the bandwidth becomes narrower and narrower until it reduces to a single point, which is equivalent to a FT frequency.

Decomposition level	Window size	Frequency band
1	2	512–1,023
2	4	256–511
3	8	128–255
4	16	64–127
5	32	32–63
6	64	16–31
7	128	8–15
8	256	4–7
9	512	2–3
10	1,024	1

The above table can be illustrated using a time–frequency plane. Figure 3.1 shows the time–frequency planes of both DWT and STFT side by side. The wavelet time–frequency plane is shown on Fig. 3.1 Left. In contrast, STFT uses one window size for all frequencies as shown in Fig. 3.1 Right. Compared with STFT, a

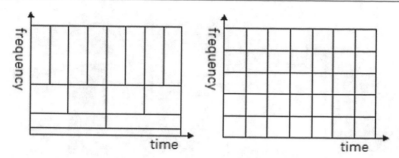

Fig. 3.1 Different frequency tiling of spectral plane. Left: the wavelet spectrum; Right: the STFT spectrum

DWT yields very high resolution at lower frequencies while sacrifices resolution at higher frequencies. Therefore, a DWT is a multiresolution tool.

Wavelets analyze and represent a signal with multiresolution. This is extremely useful, because lower resolution represents a summary and higher resolution represents fine details of a signal, both are essential in signal analysis and representation. The multiresolution representation is done through repeating rounds of scaling (low pass, L) and wavelet transform (high pass H) on a signal. The scaling captures the low frequency information of the signal and the wavelet captures the high frequency information of the signal. At each round of the wavelet transform, a low-resolution signal (low frequency L) and a fine details signal (high frequency H) are obtained, both are half the size of the original signal. Since the information in the fine details signal is usually scarce, most of the information in the original signal is captured by the lower resolution version, this achieves great efficiency of signal representation. To recover the signal, the inverse wavelet transform is applied, which is done through repeating rounds of expansion.

3.3   Fast Wavelet Transform

3.3.1   DTW Decomposition Tree

For a 2D image, the rows and columns are treated as 1D signals. Due to the separability property of DWT, the two passes at each round of the DWT are done at the rows and the columns separately. The 2D digital wavelet transform on an image is illustrated in Figs. 3.2, 3.3 and 3.4.

- **Horizontal transform**. At the first step of level 1 decomposition, each row of the image is scaled (weighted average) and wavelet transformed (weighted difference).

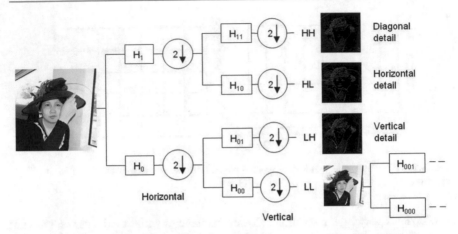

Fig. 3.2 The 2D DWT decomposition tree for a lady image

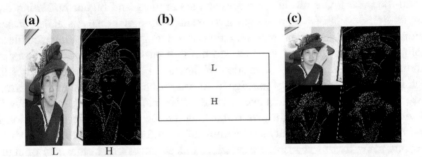

Fig. 3.3 Illustration of 2D DWT process on a lady image. **a** Horizontal transform; **b** vertical transform; **c** spectrum of level 1 2D DWT decomposition

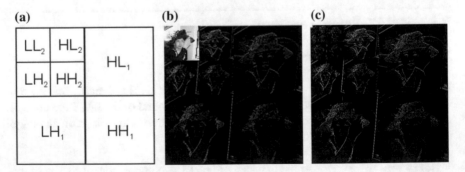

Fig. 3.4 Two levels of 2D wavelet decomposition. **a** The spectrum plane of two levels of 2D DWT; **b** the spectrum of two levels of 2D DWT on the lady image; **c** the complete decomposition of the lady image

– The result from the first step are two half images, one with scaling coefficients (L) and the other with wavelet coefficients (H), both are half width of the image row (Fig. 3.3a).

- **Vertical transform**. In the next, the scaling (L) and wavelet transform (H) are applied on each column of the two half images from the previous step (Fig. 3.3b).

  – The result from the second step is four quarter sized images, which are: summary (LL, top left), vertical details (LH, bottom left), horizontal details (HL, top right), and diagonal details (HH, bottom right) (Fig. 3.3c).

- **Level 2 decomposition**. The above steps are repeated on the LL image for the next round of DWT (Fig. 3.4b).

- **DWT spectrum**. The level 1 decomposition process can be repeated until the summary image can no longer be decomposed further, the final spectrum of the wavelet transform on the lady image is shown in Fig. 3.4c. It can be observed that the DWT spectrum captures the essence of the image while discarding all the redundant details. This is very useful for image analysis.

This process can be summarized in mathematical terms. Suppose the scaling and wavelet functions are ϕ, ψ, respectively. At each level of decomposition, the following 4 quarter sized images are resulted from the DWT by using (3.5): average/summary image $\phi(x, y)$, horizontal difference/details image $\psi^H(x, y)$, vertical difference/details image $\psi^V(x, y)$, and diagonal difference/details image $\psi^D(x, y)$.

$$\phi(x, y) = \phi(x)\,\phi(y) \quad \rightarrow \text{LL} \tag{3.6}$$

$$\psi^H(x, y) = \phi(x)\,\psi(y) \quad \rightarrow \text{HL} \tag{3.7}$$

$$\psi^V(x, y) = \psi(x)\,\phi(y) \quad \rightarrow \text{LH} \tag{3.8}$$

$$\psi^D(x, y) = \psi(x)\,\psi(y) \quad \rightarrow \text{HH} \tag{3.9}$$

3.3.2 1D Haar Wavelet Transform

The DWT can be demonstrated using Haar wavelet, the scaling function and wavelet function of Haar transform are shown in Fig. 3.5. Given the unique characteristic or shape of the Haar transform functions, the scaling and wavelet transform of Haar wavelet become simply the *average* and *difference* (or *details*).

Suppose x and y are two neighboring points, the scaling coefficient and wavelet coefficient are given by s and d, respectively:

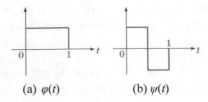

(a) $\varphi(t)$ (b) $\psi(t)$

Fig. 3.5 Harr scaling function and wavelet function

$$s = (x+y)/2 \quad \text{and} \quad d = (x-y)/2 \tag{3.10}$$

The inverse Haar transform is then given by *addition* and *subtraction*:

$$x = s+d \qquad \text{and} \quad y = s-d \tag{3.11}$$

- Given an even length discrete signal of $(a_0, a_1, \ldots, a_{2n}, a_{2n+1})$.
- It is first organized into pairs $((a_0, a_1), \ldots, (a_{2n}, a_{2n+1}))$.
- By applying the first round of Haar transform, the coefficients of the transform are given by $((s_0, s_1, \ldots, s_n), (d_0, d_1, \ldots, d_n))$.
- The second round of Haar transform can be performed on the sequence of s, and so on so forth.

For example, suppose [11, 9, 5, 7] is a 4-point digital signal, the following demonstrates the process of a Haar wavelet transform on the signal.

Resolution	Averages	Details
4	[11, 9, 5, 7]	
2	[10, 6]	[1, −1]
1	[8]	[2]

Therefore, the Haar wavelet transform of [11, 9, 5, 7] is given by [8, 2, 1, −1]. As can be seen, after the wavelet transform, the first value captures the most significant information while the last two values are very small. This is helpful in signal processing and analysis, because more attention can be given to the most significant information.

The wavelet transform can be performed more efficiently by using matrix multiplication. The following is an example of 4 × 4 Haar wavelet transform matrix.

$$H_4 = \frac{1}{4} \begin{bmatrix} 1 & 1 & 1 & 1 \\ 1 & 1 & -1 & -1 \\ 2 & -2 & 0 & 0 \\ 0 & 0 & 2 & -2 \end{bmatrix} \tag{3.12}$$

For the above 4-point signal, using the Haar wavelet transform matrix, the transform coefficients are given by

$$h_4 = \frac{1}{4} \begin{bmatrix} 1 & 1 & 1 & 1 \\ 1 & 1 & -1 & -1 \\ 2 & -2 & 0 & 0 \\ 0 & 0 & 2 & -2 \end{bmatrix} \begin{bmatrix} 11 \\ 9 \\ 5 \\ 7 \end{bmatrix} = \begin{bmatrix} 8 \\ 2 \\ 1 \\ -1 \end{bmatrix} \tag{3.13}$$

3.3.3 2D Haar Wavelet Transform

For a 2D image, this is done on both the rows and columns separately.

Suppose the following is a 4 × 4 image I:

102	56	68	152
24	62	46	32
52	92	72	84
76	60	92	60

Step 1. *Horizontal scaling* of image I (horizontal pairwise average, L):

79	110
43	39
72	78
68	76

Step 2. *Horizontal wavelet transform* of I (horizontal pairwise difference, H):

23	−42
−19	7
−20	−6
8	16

The image I_c after horizontal transform by combining the results from the above two steps:

79	110	23	−42
43	39	−19	7
72	78	−20	−6
68	76	8	16

Step 3. *Vertical scaling* of I_c (vertical pairwise average, LL, HL)

61	74.5	2	−17.5
70	77	−6	5

Step 4. *Vertical wavelet transform* of I_c (vertical pairwise difference, LH, HH)

18	35.5	21	−24.5
2	1	−14	−11

The image after the first round of Haar wavelet transform by combining the results from Step 3 and 4:

LL	61	74.5	2	−17.5	HL
	70	77	−6	5	
LH	18	35.5	21	−24.5	HH
	2	1	−14	−11	

The second round of Haar DWT repeats the Steps 1–4 on the LL band and this process can be continued until required levels of decomposition is achieved. Similar to the 1D case, the first quarter of the wavelet transformed image contains the most significant information.

3.3.4 Application of DWT on Image

Figure 3.6 demonstrates the complete process of computing the Haar wavelet transform on the lady image using the DWT decomposition tree of Fig. 3.2. At each next level of the decomposition, the image is halved, therefore, the DWT is very efficient and fast.

3.4 Summary

Wavelets are an extension or an improvement to windowed FT in two aspects: *orthogonality* and *multiresolution*. Orthogonality means that there is no redundancy between DWT channels. Multiresolution means to analyze an image by zooming in and zooming out, which is like studying a map with a magnifying glass. This is achieved by adapting image resolution to wavelet size/scale.

The contrast between wavelets and windowed FT can be easily understood in frequency plane as shown in Fig. 3.1. Basically, with DWT, we retain higher resolution at very low-frequency band at the cost of losing resolution at high-frequency band. This is sensible because low-frequency information is much more important than high frequency information to human perception.

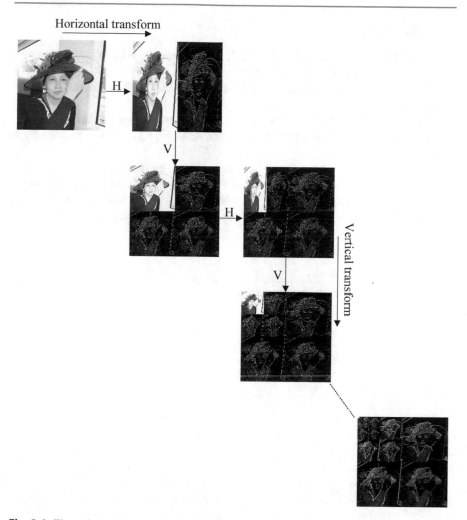

Fig. 3.6 Illustration of the computing process of Haar wavelet transform

The computation of 2D DWT is very efficient due to the separability property of WT. In practice, the computation of 2D DWT follows a decomposition tree or algorithm as shown in Fig. 3.2. Generally, there needs just a few rounds of repeat DWT decomposition to transform an image. In terms of understanding the DWT computation process, it is important to understand how Haar wavelet transform works. Once the process of Haar wavelet transform is understood, one can just replace the Haar wavelet with any other wavelet.

3.5   Exercises

1. Given a signal $f = (5, 8, 3, -4, 7, 8, 5, 3)$, compute the Haar wavelet transform of the signal and give all coefficients s and d.
2. Plot all wavelet basis functions ψ_s for all valid scales of the signal computed in problem 1.
3. What is the mathematical concept behind the computation of the Haar coefficients (hint: consider the relationship between Haar, FT, and STFT) and why is it important?
4. Choose an image of your own and apply the Haar wavelet transform to the image using the Matlab code shown in this web page: https://au.mathworks.com/help/wavelet/ref/dwt2.html. Try different images and write a short report on your discovery, e.g., what kind of features it has captured and why.

Part II

Image Representation and Feature Extraction

Sort the wheat from the chaff.

Introduction

A digital color image $I = \{p_{ij}\}$ is a matrix of pixels, each pixel is a three-dimensional color vector $p_{ij} = (r_{ij}, g_{ij}, b_{ij})$, representing the three color components of the pixel. For example, an image with $m \times n$ pixels is represented as an $m \times n$ matrix:

$$
\begin{aligned}
I &= \begin{bmatrix} p_{11}, p_{12}, \ldots, p_{1n} \\ p_{21}, p_{22}, \ldots, p_{2n} \\ \ldots \\ p_{m1}, p_{m2}, \ldots, p_{mn} \end{bmatrix} \\
&= \begin{bmatrix} (r_{11}, g_{11}, b_{11}), (r_{12}, g_{12}, b_{12}), \ldots, (r_{1n}, g_{1n}, b_{1n}) \\ (r_{21}, g_{21}, b_{21}), (r_{22}, g_{22}, b_{22}), \ldots, (r_{2n}, g_{2n}, b_{2n}) \\ \ldots \\ (r_{m1}, g_{m1}, b_{m1}), (r_{m2}, g_{m2}, b_{m2}), \ldots, (r_{mn}, g_{mn}, b_{mn}) \end{bmatrix}
\end{aligned}
\tag{II.1}
$$

For color processing and analysis, a color image is often represented as a composition of three component color images or image planes: $I = [I_r, I_g, I_b]$:

$$
I_r = \begin{bmatrix} r_{11}, r_{12}, \ldots, r_{1n} \\ r_{21}, r_{22}, \ldots, r_{2n} \\ \ldots \\ r_{m1}, r_{m2}, \ldots, r_{mn} \end{bmatrix} \quad I_g = \begin{bmatrix} g_{11}, g_{12}, \ldots, g_{1n} \\ g_{21}, g_{22}, \ldots, g_{2n} \\ \ldots \\ g_{m1}, g_{m2}, \ldots, g_{mn} \end{bmatrix} \quad I_b = \begin{bmatrix} b_{11}, b_{12}, \ldots, b_{1n} \\ b_{21}, b_{22}, \ldots, b_{2n} \\ \ldots \\ b_{m1}, b_{m2}, \ldots, b_{mn} \end{bmatrix} \tag{II.2}
$$

To extract texture features from a color image, it is usually converted to a gray level image. A gray level image is basically a 3D function $z = g(x, y)$, where (x, y) is the pixel position and z is the intensity. By modeling an image as a mathematical function, advanced theories on mathematical analysis, algebra, and statistics can be employed for image analysis.

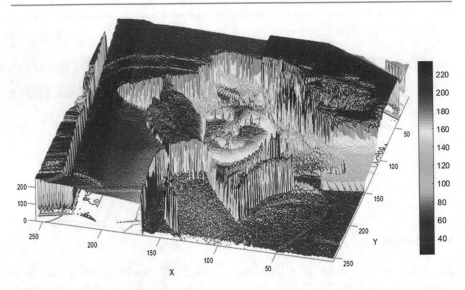

Fig. II.1 The 3D map for a 2D image

Figure II.1 shows the 3D function for the lady image in Fig. 3.2, where the colors in the 3D space represent the intensity of the image pixels. It is observed that a gray level image is essentially a 3D terrain or a 3D surface in space, this modeling is helpful because advanced geometric theories can also be employed on texture analysis such as *fractal*.

Images are a powerful media; in fact, human beings learn to understand the world visually before learning to read and write. It is widely known that a picture is worth a thousand of words. This is because people read images much more efficiently and effectively than reading the text. Indeed, people nowadays get more information from TV, videos, and cinema than from books.

Although images are a powerful media, it is much more difficult to mine semantic information from images than from textual documents, due to each image has a large number of image pixels and there is no visual dictionary to represent images. Image size on Internet nowadays ranges from several hundreds of thousands of pixels to several millions of pixels and it is even bigger in private storage. Image size is going to get bigger and bigger as the image capturing devices get more and more powerful and the Internet speed gets faster and faster.

To make the mining task even more complicated, an image is usually corrupted or degraded. An observed image $I(x, y)$ is typically a degraded version of an ideal image $f(x, y)$. In general, the relation between $f(x, y)$ and $I(x, y)$ is governed by $I(x, y) = T_g [T_d (f)]$, where T_d and T_g are color and geometric *degradation operators*, respectively. In practice, T_d is usually modeled as a convolution while T_g is typically modeled as a spatial transform such as affine transform or similarity transform. T_d and T_g are usually unknown. The goal is to model the unknown image

$f(x, y)$ through the observed image $I(x, y)$ by the means of some a priori information or assumption.

It is known the world is consisted of objects and each object can be described with a small number of properties or features. For example, a human being can be described by his/her gender, height, weight, hair, complexion, dressing, ethnicity, etc. Similar to interpreting the world, human beings tend to group image pixels into objects and describe them accordingly when reading an image.

The idea of image analysis or image data mining is to mine a small number of features from the large number of image pixels, so that images can be described effectively and similar images can be classified. Ideally, the features should be semantic so that human beings can understand them. However, to mine semantic features like water, sky, horse, tree, etc., is extremely difficult for machine. Therefore, a typical image analysis mechanism starts from low-level features such as color, shape, texture, etc. Once low-level features are extracted, higher level features can be learned through machine learning methods which will be described in Part III, and similar images can be retrieved through image retrieval methods which will be described in Part IV.

In Part II, the three types of common low-level image features will be described in detail.

Color Feature Extraction

4

Every picture tells a story, by colors.

4.1 Introduction

Arguably, color is the most important feature of an image. After all, people see this world as colors or the world presents itself to us as colors. However, color is a complex topic and difficult to understand. As a matter of fact, few people are good at painting a picture or image. There are infinite number of colors in this world and colors can be created from different types of palettes. Computers use a trichromatic palette to mix all the colors in this world. That means each color in computers is represented as a three-dimensional vector (c_1, c_2, c_3). These color vectors created a 3D color space. Depending on how each of the trichromatic colors is defined, different color spaces or color models have been created.

The most commonly used color space is the RGB color space, where each of the colors is defined by adding three primary colors in the visible light spectrum (red, green, and blue) with various proportions. Other commonly used color spaces include LUV, HSV/HSL/HSI, YCrCb.

Color spaces are models for the representation of pixel values. To compare and classify color images, however, we need to analyze and understand the *color patterns* in an image. In order to understand the color patterns in an image, we extract color features from the image and compare them with features of other images. Color features are usually based on color statistics computed from an image or regions of an image.

A number of color features have been proposed in the literature including color *histogram*, *color moments* (CM), *color coherence vector* (CCV), color *correlograms*, etc. MPEG-7 also standardizes a number of color features including *dominant color descriptor* (DCD), *color layout descriptor* (CLD), *color structure descriptor* (CSD), and *scalable color descriptor* (SCD).

© Springer Nature Switzerland AG 2019
D. Zhang, *Fundamentals of Image Data Mining*, Texts in Computer
Science, https://doi.org/10.1007/978-3-030-17989-2_4

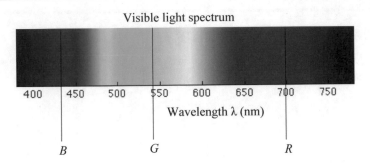

Fig. 4.1 Visible light spectrum and the tristimulus

4.2   Color Space

To process and analyze color images, we need to understand how different color models work, and their applications in image processing and analysis. Color is a complex theory, there are infinite number of colors in this world and colors can be created using a variety of ways. There needs a standard color model so that colors can be reproduced with accuracy and colors produced in different applications by different devices can be translated interchangeably. The first step is to find a way of representing each color numerically and identify the space or gamut of all visible colors.

The building of a standard color model is made possible thanks to the three scientists: *Isaac Newton, James Clerk Maxwell,* and *Hermann Grassmann. Newton* laid the foundation of our understanding of colors by first splitting a light source into spectral or pure colors (rainbow colors, Fig. 4.1). This lets us to understand that a light is a mixture of pure colors and colors are just reflectance of lights of different wavelengths by objects. *Maxwell* found that by projecting and superimposing the three red, green, and blue monochromatic pictures on the screen, other colors in the scene such as orange, yellow, purples, etc., also showed up, suggesting other colors can be created by mixing red, green, and blue colors. *Grassmann* found that colors are additive. That means any color can be matched by a linear combination of three other colors (primaries), provided that none of those three can be matched by a combination of the other two; and a mixture of any two colors can be matched by linearly adding together their primary components.

4.2.1   CIE XYZ, *xyY* Color Spaces

Modern color models are all based on XYZ color space created by CIE (International Commission on Illumination) in 1931. The CIE XYZ color space was created using Maxwell's *tristimulus* theory, which is based on the theory of *trichromatic* color vision found by Young and Helmholtz [1], who discovered that human vision consists of three types of cones, which are sensitive or responsive to three narrow

Fig. 4.2 CIEXYZ color matching functions of human vision

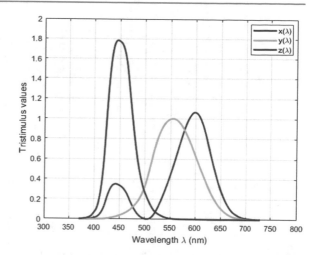

bands of visible lights. Figure 4.2 shows the three color matching functions which indicate human eyes' response to visible colors. It can be observed that three functions peak at around 600 nm, 550 nm, and 450 nm, respectively. It demonstrates that human vision is most sensitive to three bands of lights, which are perceived as red, green, and blue.

Based on this discovery, CIE uses three *primary colors* to match out all the *spectral colors*, i.e., pure colors or colors with a single wavelength (Fig. 4.1) [2]. The three primary colors are all pure colors, they are R (700 nm), G (546.1 nm), and B (435.8 nm), respectively, which are shown on the visible color spectrum in Fig. 4.1. The choice of the three particular primaries was due to practical reason at that time. The primaries G (546.1 nm) and B (435.8 nm) were chosen because they could easily be reproduced using mercury excitation lines. The 700 nm primary color was chosen is because the hue near that wavelength is homogenous and nearly constant, therefore, slight inaccuracy in production of the wavelength of this spectral primary would introduce no error at all.

The three primary stimulus are projected on a screen with relative power and are mixed/added by various proportions to match out each of the *spectral colors* in the visible color spectrum using Grassmann's laws. Each color can now be represented as a three-value tuple (R, G, B). The R, G, B values are then normalized using the formulas in (4.1) to remove the intensity from the color representation. The normalised r, g, b values are purely chromatic values. This creates three color matching functions (CMF) $r(\lambda)$, $g(\lambda)$, and $b(\lambda)$, where λ is the wavelength. The color space created based on the three CMFs is called CIE RGB color space.

$$\left. \begin{array}{l} r = \frac{R}{R+G+B} \\ g = \frac{G}{R+G+B} \\ b = \frac{B}{R+G+B} \end{array} \right\} \tag{4.1}$$

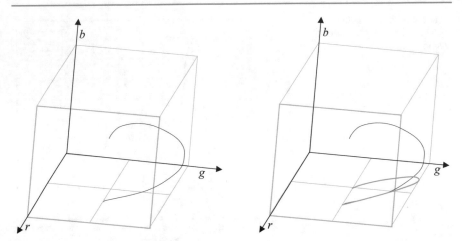

Fig. 4.3 Color matching function of spectral colors. Left: *rgb* curve of spectral colors; Right: projection of the *rgb* curve onto the 2D *rg* plane (cyan)

Now if we plot the (r, g, b) coordinates of all the spectral colors in 3D space, it forms a curve (Fig. 4.3 Left) [3]. It is easy to see from (4.1) that $r + g + b = 1$ or $b = 1 - r - g$, which means b is a dependent function of r and g, so there is no need to keep the information b. Therefore, by projecting the *rgb* curve into the 2D *rg* plane, we get the horseshoe-shaped 2D *rg* curve which is shown as cyan color in Fig. 4.3 Right.

This *rg* curve is then transformed to the CIE *xy* curve by aligning the $g(\lambda)$ with the CIE luminosity function $V(\lambda)$ and removing the negative values in $r(\lambda)$. The color space created based on the *xy* curve is called CIE XYZ color space.

The colors on the 2D *xy* curve are all *spectral colors*, to obtain *nonspectral colors*, i.e., mixed colors or colors with multiple wavelengths, we draw a straight line between any two points on the *xy* curve. Then each point on the straight line represents a nonspectral color mixed by the two colors at both ends of the line according to Grassmann's second law. For example, by connecting the two primary colors R (700 nm) and B (435.8 nm) on the *xy* curve, purple colors are created. By this way, all possible *nonspectral colors* can be created. In practice, a white color or white point is first defined, such as D65 which represents the midday Sunlight, and lines are drawn from the white point to each of the spectral colors on the *xy* curve. Then colors of different purity are created by mixing the white color with each of the pure colors on the curve. Figure 4.4 shows the color gamut of CIE XYZ color space or CIE *xy* gamut [2]. The color gamut created in this way is a *hue* and *saturation* gamut.

Fig. 4.4 CIE *xy* color gamut

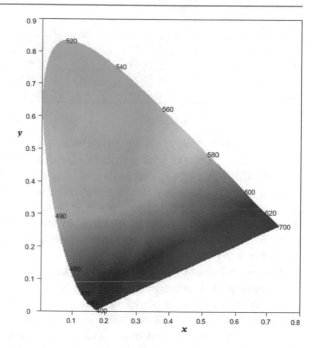

CIE *xy* gamut is a chromaticity domain, it does not specify luminance or brightness of colors. To create object colors, the luminance/brightness must be given as the third dimension, named as *Y*. Therefore, CIE *xyY* color space is created and is called the *object color space*, where *x* and *y* are chromaticity values, and *Y* is the luminance value.

CIE XYZ color space is a cornerstone for modern color modeling. The significance of CIE XYZ color space can be summarized in the following:

- Provides a color gamut with all possible colors,
- Specifies each color with a three-value tuple or a 3D vector (x, y, z),
- Provides a reference for all other color models,
- It is a device-independent color space.

4.2.2   RGB Color Space

Digital images are generated using RGB colors. RGB colors are device dependent, which means that each type of digital devices typically uses a different set of RGB primaries to generate colors. For example, computers use a standard RGB color model or sRGB, which is based on the following three primaries chosen from the CIE XYZ color gamut.

Chromaticity	x	y	Y
R	0.6400	0.3300	0.2126
G	0.3000	0.6000	0.7125
B	0.1500	0.0600	0.3290
W(hite)	0.3127	0.3290	1.0000

The three primaries of sRGB color model are shown in Fig. 4.5 [4]. The gamut of sRGB color space is a triangle inside the CIE XYZ gamut.

The RGB color gamut can be regarded as a color palette. All possible colors created by the RGB palette can be visualized in a 3D cube, called RGB color space as shown in Fig. 4.6. The colors in the RGB color cube are usually quantized for the convenience of viewing. Given the pixel values (r, g, b) from a color image, a color c can be defined in the RGB color space and reproduced by mixing the three primaries using $c = rR + gG + bB$.

It is clear that the gamut of any RGB color model is a triangle inside the CIE XYZ color gamut, consequently, a RGB color space cannot represent all visible colors. In case of a color out of the RGB color gamut, an approximation has to be made. For a color C out of the RGB gamut, the approximation of C is given by the color C_A at the intersection of the RGB triangle and line CW which is the connection between color C and white point W. The approximation is usually acceptable because C_A is just a desaturated color from C.

Fig. 4.5 The sRGB triangle gamut shown inside the CIE XYZ gamut

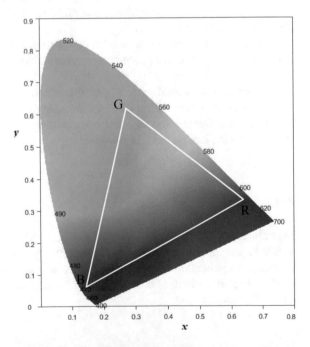

Fig. 4.6 RGB color space in 3D

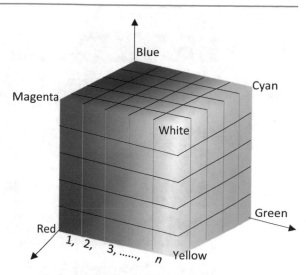

RGB module is useful for color display and printing, however, it is not desirable for image processing and analysis. This is because the three channels are dependent on each other and there is a high correlation between the three channels, which means, change any one of the color channels will change the other two color channels. Furthermore, RGB color space is not perceptually uniform, meaning that the same amount of numerical change in color values does not correspond to about the same amount of visually perceived change. This leads to color spaces with separation of luminance from chromaticity and color spaces with uniform color distance.

4.2.3   HSV, HSL and HSI Color Spaces

RGB color model is efficient because it just uses three primaries to create all required colors. However, the RGB color model is not intuitive because it does not conform to how human beings understand and make colors. For example, artists and painters do not use RGB mixture to make colors, instead, they use pigments to mix with either white or black or both (gray) to make required colors. The pigments are equivalent to pure colors or spectral colors, when they are mixed with white or black, lighter or darker colors are created; when they are mixed with gray, colors with different purity or saturation are created. Figure 4.7a shows how different tints, shades, and tones of reddish colors are created by artists.

The way artists and painters making colors is the idea behind the HSV color model. It demonstrates that a color can be specified by three components/properties: Hue, Saturation, and Value or (H, S, V), where Hue is a pure color and Value = Brightness. The *Hue* tells what color it is, it is determined by the dominant wavelength on the visible color spectrum (Fig. 4.1). The *Saturation* tells how much or how colorful is the color, the more saturated the color, the more vibrant or vivid

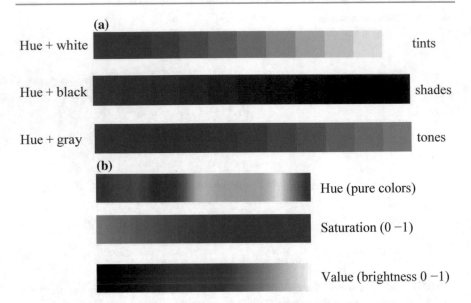

Hue + white   tints

Hue + black   shades

Hue + gray   tones

Hue (pure colors)

Saturation (0 −1)

Value (brightness 0 −1)

Fig. 4.7 **a** Artists' way of making reddish colors. **b** Components of HSV

the color is. The *Value* tells how bright or dark is the color, colors become disappearing when they are too dark or too bright. Therefore, a color specified by the (*h*, *s*, *v*) values makes a lot more sense than that given by the (*r*, *g*, *b*) values. Multimedia editors and image processing software nowadays all provide intuitive HSV color picker simply because users have better chance to make the desired colors using HSV model than using other color models.

Figure 4.7b demonstrate the HSV color making using red colors as an example. The first bar shows all the pure colors or hues; the second bar shows red colors with a different purity or saturation (but with the same brightness); the third bar shows red colors with different brightness.

To create the HSV color model, pure colors (spectral colors) are first collected and put on a circle or a ring (Fig. 4.8 Left), colors with different saturation are created along the radii of the circle to create a hue–saturation disk/wheel (Fig. 4.8 Right). Hue–saturation disks with different brightness are then generated and stacked on top of each other to make a color cylinder which is the HSV color space, shown in Fig. 4.9a. For HSV, the most saturated colors are on the top of the cylinder and the top of the cylinder has a V value of 1. For HSL, the most saturated colors are in the middle of the cylinder and the top of the cylinder is the white color (L = 1), this is shown in Fig. 4.9b.

It is observed from the Value strip of Fig. 4.7b that as colors become darker or brighter, they become less colored, consequently, as shown in Fig. 4.9a, b, colors on the HSV and HSL cylinders become more and more redundant as they go down the cylinders and as they go up the HSL cylinder. Therefore, the actual HSV color space is often shown as a single cone (Fig. 4.9c) while HSL color spaces is often

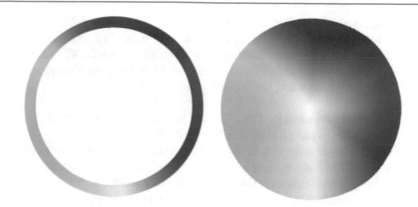

Fig. 4.8 Hue and saturation. Left: pure colors on a ring; Right: hue-saturation wheel

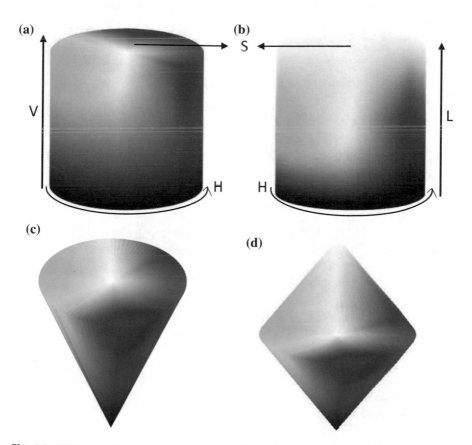

Fig. 4.9 HSV and HSL color spaces. **a** HSV cylinder; **b** HSL cylinder; **c** HSV cone; **d** HSL double cone

shown as a double cone (Fig. 4.9d). Because every slice of the HSV/HSL cone is colorful, the radiuses of the cone are called *chroma* instead of *saturation*.

Digital images are represented using RGB colors. To derive HSV, HSL and HSI color values, RGB values are normalized into 0–1. Given a (R, G, B) color, the H, S, V values are computed using the following guidelines.

- The Hue values are organized into 0°–360° around a pure color circle, with red color at 0°/360°, green at 120°, and blue at 240°;
- To determine the H value of a (R, G, B) color, the maximum of the three values is used to determine the *dominant hue* on the pure color circle and the difference of the other two values tells what side is the RGB color located at the *dominant hue*;
- The Value/Intensity/Lightness of the (R, G, B) color is determined by either the maximum of the three RGB values or the average or in between;
- The Saturation of a color is determined by how far the color is from the pure color circle (Hue) which has a color saturation of 1. Saturation is given in percentage, e.g., 40%.

Let

$$M = max(R, G, B)$$
$$m = min(R, G, B)$$
$$C = M - m$$

The maximum of the three RGB channels M dominants the *hue* and *brightness* of a color. The C value, called *chroma*, is proportional to *saturation* of a color. With these in mind, HSV values can be computed using the following formulas:

$$H = \begin{cases} 0 & \text{if } M = m \\ 60° \times \frac{G-B}{C} + 0°, & \text{if } M = R \text{ and } G \geq B \\ 60° \times \frac{G-B}{C} + 360°, & \text{if } M = R \text{ and } G < B \\ 60° \times \frac{B-R}{C} + 120°, & \text{if } M = G \\ 60° \times \frac{R-G}{C} + 240°, & \text{if } M = B \end{cases} \qquad (4.2)$$

$$S = \begin{cases} 0 & \text{if } M = 0 \\ \frac{C}{M} = 1 - \frac{m}{M}, & \text{otherwise} \end{cases} \qquad (4.3)$$

$$V = M \qquad (4.4)$$

For *HSL* model:
H is the same as (4.2)

$$L = \frac{1}{2}(M + m) \qquad (4.5)$$

$$S = \begin{cases} 0 & if\ M = m \\ \frac{C}{2L}, & if\ L \leq \frac{1}{2} \\ \frac{C}{2-2L}, & if\ L > \frac{1}{2} \end{cases} \qquad (4.6)$$

For *HSI* model:

H is the same as (4.2)

$$I = \frac{1}{3}(R + G + B) \qquad (4.7)$$

$$S = \begin{cases} 0 & if\ I = 0 \\ \frac{I-m}{I} = 1 - \frac{m}{I}, & otherwise \end{cases} \qquad (4.8)$$

HSV values can also be derived using the following formulas:

$$H = \arctan\left(\frac{\beta}{\alpha}\right) \qquad (4.9)$$

$$S = \sqrt{\alpha^2 + \beta^2} \qquad (4.10)$$

$$V = max(R, G, B) \qquad (4.11)$$

where

$$\alpha = R - \frac{1}{2}(G + B), \quad \beta - \frac{\sqrt{3}}{2}(G - B) \qquad (4.12)$$

The *H* and *S* components are invariant to lighting variations or intensity changes. Intensity changes are only reflected in the *V* component, which can be corrected by a linear scaling.

Figure 4.10 shows a color image and its H, S, V channels. In the H image, white and black are starting and arrival points on the color wheel, they represent Red color and Yellow color. Gray intermediate levels are corresponding to intermediate hues on the wheel. Both the S and V channels are in the 0–1 range. For S channel, white is pure color and black is minimum saturation. For V channel, white is very bright and black otherwise.

It can be observed from the S and V channels that both the yellow plants at the bottom left of the image are highly saturated and also very bright, while the red leaf tree is highly saturated but with moderate brightness. Notice the shadow at the bottom left of the color image has no to little effect on the H and S channels, it only affects the V channel.

Fig. 4.10 A color image on the left and its H, S, V channels on the right columns

4.2.4   CIE LUV Color Space

The CIE XYZ color space is nonuniform in terms of color differences, so is RGB color space. Nonuniform means that the calculated difference between the two colors is not proportional to their perceived color difference. This phenomenon is shown as MacAdam ellipses in the CIE xy gamut in Fig. 4.11. Each ellipse shown in the figure represents colors within the just-noticeable-difference (JND) threshold. In other words, colors within each ellipse are perceivably the same. As can be seen, the sizes of the ellipses in different areas of the gamut vary significantly. This implicates that colors within certain distance may be perceived as the same color in one area but as different colors in another area. This causes confusions for many

Fig. 4.11 MacAdam ellipses (magnified 10 times) on the CIE xy gamut

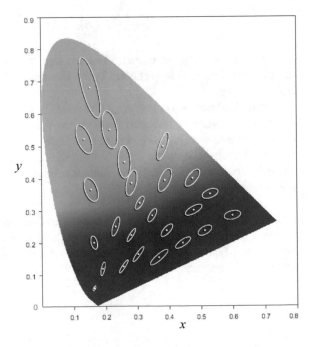

color applications including displaying, image processing, and image analysis, which rely on computing color differences.

To overcome the nonuniform color spread problem, CIE Luv (1960) and CIE Lu'v' (1976) color spaces have been created. Both Luv and Lu'v' spaces are transformed from CIE XYZ space, and the two color spaces are only different at the v component. The idea is to stretch or squeeze the CIE xy gamut at certain directions so that the MacAdam ellipses are made to equal size.

$$L^* = \begin{cases} 116\left(\frac{Y}{Y_n}\right)^{\frac{1}{3}} - 16 & if \frac{Y}{Y_n} > 0.008856 \\ 903.3\left(\frac{Y}{Y_n}\right) & if \frac{Y}{Y_n} \leq 0.008856 \end{cases} \tag{4.13}$$

$$u = \frac{4X}{X + 15Y + 3Z} = \frac{4x}{3 - 2x + 12y} \tag{4.14}$$

$$v = \frac{6Y}{X + 15Y + 3Z} = \frac{6y}{3 - 2x + 12y} \tag{4.15}$$

$$u' = u, \quad v' = 1.5v \tag{4.16}$$

where Y_n is the luminance of the white point and

$$\left. \begin{array}{l} x = \frac{X}{X+Y+Z} \\ y = \frac{Y}{X+Y+Z} \end{array} \right\} \tag{4.17}$$

L^* scales from 0 to 100 due to the relative luminance (Y/Y_n) scales from 0 to 1. The cubic root function of L^* is nonlinear and is intended to mimic the *logarithmic response* of human eyes to lightness. The transformed uv and $u'v'$ gamuts marked with the MacAdam ellipses are shown in (4.12). It can be seen that the differences between the sizes of the ellipses are considerably reduced (refer to Fig. 4.11) (Fig. 4.12).

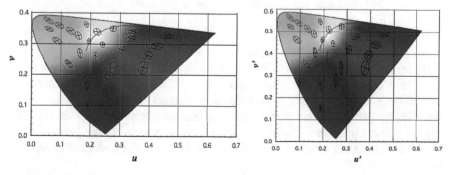

Fig. 4.12 CIE Luv and Lu'v'. Left: MacAdam ellipses on CIE uv gamut; Right: MacAdam ellipses on CIE $u'v'$ gamut

The nonuniformity can be further reduced by using the u^*v^* chromaticity:

$$u^* = 13L^*\left(u' - u'_n\right) \tag{4.18}$$

$$v^* = 13L^*\left(v' - v'_n\right) \tag{4.19}$$

where (u'_n, v'_n) are the coordinates of the white point on the (u', v') gamut.

To derive LUV from RGB color space, RGB values are first transformed into XYZ values using the following matrix, where RGB values have been normalized to 0–1.

$$\begin{bmatrix} X \\ Y \\ Z \end{bmatrix} = \begin{bmatrix} 0.4124 & 0.3576 & 0.1804 \\ 0.2126 & 0.7152 & 0.0722 \\ 0.0193 & 0.1192 & 0.9503 \end{bmatrix} \begin{bmatrix} R \\ G \\ B \end{bmatrix} \tag{4.20}$$

Although CIE LUV color spaces are close to uniform color spaces, the source RGB primaries are assumed to be known, so that a specific transform matrix of (4.20) can be used. This can be an issue for image applications, because the source (device) RGB primaries are usually unknown.

4.2.5   Y′CbCr Color Space

Both HSV and LUV color spaces are based on the same idea, i.e., the separation of luminance from chromaticity. This idea has been found ideal and desirable for most of the color applications including image processing and feature extraction. Y′CbCr is another such kind of color space, which is often used for image compression and representation. The transformation from RGB space to Y′CbCr space is given by the following equation:

$$\begin{bmatrix} Y' \\ Cb \\ Cr \end{bmatrix} = \begin{bmatrix} 0 \\ 128 \\ 128 \end{bmatrix} + \begin{bmatrix} 0.299 & 0.587 & 0.114 \\ -0.169 & -0.331 & 0.500 \\ 0.500 & -0.419 & -0.081 \end{bmatrix} \begin{bmatrix} R \\ G \\ B \end{bmatrix} \tag{4.21}$$

where Y' is the luminance, Cb is the blue component, and Cr is the red component. Both RGB values and Y′CbCr values are in the range of [0, 255].

By separating the luminance Y' from the chromaticity Cb and Cr, most of the image information has been concentrated onto Y'. This is ideal for image communication and representation, because each channel can now be treated independently instead of treating all the three channels equally as in the RGB space. For example, more importance can be given to Y' as in many situations. This makes the communication more efficient (e.g., fewer bits for color channels) and representation more compact.

Figure 4.13 shows an example of Y′CbCr channels from a flower image. It can be seen that the Y′ channel is basically the gray level version of the original image,

Fig. 4.13 The flower image on the leftmost and its Y', Cb, and Cr channels on the right

while the Cb and Cr channels are just for the color information, both Cb and Cr channels contain very little information compared with the Y' channel. In the TV broadcast, the Y' channel can be sent out independently to be compatible with the old noncolor TVs.

4.3   Image Clustering and Segmentation

Digital images are complex data. Unlike textual documents which are made of words from a dictionary of a small vocabulary, there is no visual dictionary or vocabulary for images. Each image consists of thousands to millions of pixels which represent color values, and the possibilities of the pixel colors are almost infinite. Therefore, the first step to analyze an image is usually to group the image pixels into a small number of regions or objects so that further analysis can be carried out, this is called *image clustering or segmentation*. There are many segmentation and clustering algorithms in the literature, in the next, we discuss two widely used algorithms in image feature extraction.

4.3.1   *K*-Means Clustering

One of the simplest segmentation methods is the *K-means* clustering. *K*-means clustering attempts to divide a dataset into K clusters with each data point belonging to the cluster with the closest mean, which serves as the centroid of the cluster. The *K*-means clustering algorithm is given as follows:

K-means (K) {
Input: X = {$\mathbf{x}_1$, $\mathbf{x}_2$, ..., $\mathbf{x}_n$}, a set of data points
Output: C = {c_1, c_2, ..., c_K}, a set of clusters

1. Randomly select K cluster centers or seeds;
2. Calculate the distance between each data point and all the K cluster means $\mathbf{m}_k$;
3. Assign the data point to the cluster with the nearest cluster mean;

4. Recalculate the new means for each cluster;
5. Repeat from step 2 until no data point needs to be reassigned.

}

K-means clustering algorithm aims at minimizing an objective function known as the *sum of squared error* or SSE function, which is given by

$$J(C) = \sum_{k=1}^{K} J_k \qquad (4.22)$$

and J_k is the SSE of the kth cluster:

$$J_k = \sum_{i=1}^{|c_k|} (\|\mathbf{x}_i - \mathbf{m}_k\|)^2 \qquad (4.23)$$

where $\mathbf{m}_k$ is the mean of cluster c_k and $| c_k |$ is the number of instances in cluster c_k.

K-means is one of the most commonly used clustering algorithms in image processing and analysis. It is especially useful for many color-based clustering such as *color quantization*, which aims to group similar colors and reduce the number of colors in an image. The key issue with a K-means clustering algorithm is the parameter K. The performance of the clustering depends on a good guess of K, however, there is no easy solution. This leads to other more sophisticated algorithms to improve the method.

4.3.2   JSEG Segmentation

JSEG method is based on the belief or assumption that color regions and textures agree with each other in an image, which means that a region with similar colors also has a similar texture. Based on this idea, the method attempts to find an agreement between the two types of features. The procedure of JSEG is summarized as follows [5]:

- **Color quantization**. At first, pixel colors of the image are quantized into a number of classes using a clustering algorithm such as K-means clustering.
- **Color map**. Pixels in the image are then replaced with the color class labels, such as 1, 2, 3, …. A class map is then formed and region growing is followed on the class map (Fig. 4.14b).
- **J-image**. The key to JSEG method is the computing of a *J-image*, which is computed by moving a local window through each pixel and calculating the SSE over the window (4.23). The SSE is related to the variance over a local neighborhood, neighborhoods with relatively uniform colors (or little to no texture) tend to have small J values while neighborhoods with high J values correspond to region boundaries or edges. The window size determines the sharpness of the J-

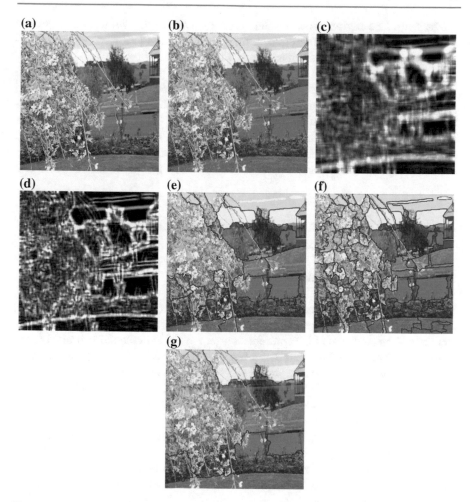

Fig. 4.14 An image segmentation using JSEG. **a** An original color image; **b** result of color quantization with 13 colors; **c** J-image at scale 3; **d** J-image at scale 2; **e** segmentation result at scale 3; **f** segmentation result at scale 2; **g** final result of segmentation after merging

image and the size of the regions that can be detected. The J-image computed using larger local window is more blurred than that computed using a smaller window (Fig. 4.14c, d).

• **Region growing**. Based on the J-image, a region growing method is carried out starting from areas with the lowest J values. After each region growing, the total J values of each region k (J_k of (4.23)) and the average J values ($\bar{J}$) of all the J_k are computed. The region growing is then repeated using a J-image with smaller scale until the $\bar{J}$ value stops decreasing (Fig. 4.14e).

- **Merging**. The region growing can result in over segmentation due to texture variations (Fig. 4.14f). Therefore, a merging process is followed by merging J-segmented regions with similar colors (Fig. 4.14g).

Due to the use of both color and texture features and a merging process, JSEG gives a less fragmented segmentation than K-means clustering. However, the performance of JSEG segmentation depends on several parameters such as the numbers of quantized colors, the seed selection threshold during the region growing, and the threshold of color similarity during the region merging. The computation is also very expensive, the segmentation of a 512×512 image can take about 4 min on a PC.

4.4   Color Feature Extraction

4.4.1   Color Histogram

The simplest feature of a color image is its histogram, which describes the color distribution within an image. To create a histogram for an image, a number of bins (N) are first created, each bin represents a group of similar colors. Each pixel in the image is then examined and put into a bin with similar colors to the pixel. After all pixels in the image are checked, the pixels in each bin are counted and each bin is represented as a value which is the number of pixels in the bin. A bar chart consists of all the N bin values is created and it's called a color histogram, which is a sequence of (c, n) pairs shown in a graph, where c is the color of the bin and n is the number of pixels in the bin. Figure 4.15 shows the three color histograms of the R, G, B channels of the Lena image in one graph.

In terms of histogram feature extraction, there are three ways to create a histogram for a color image: *component histogram*, *indexed color histogram*, and *dominant color histogram*.

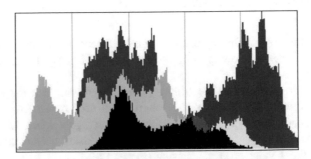

Fig. 4.15 RGB histograms of the Lena color image. Non-RGB colors are the areas of overlap between the R, G, B channels. Each channel is quantized into 256 colors or bins, which are on the horizontal axis, vertical axis shows the number of pixels in each bin or color

4.4.1.1   Component Histogram

The first way of color histogram is to split a color image into individual R, G, B channels (Fig. 4.16 top row), each individual channel is equivalent to a gray level image (Fig. 4.16 bottom row). A contrast with Fig. 4.13 tells that there is a lot of redundancy or correlation between R, G, and B channels. A histogram is first created for each individual channel. The three individual channel histograms can then be concatenated into a single histogram. If each individual color channel is represented by l, m, and n bins, respectively, the final histogram will have $N = l + m + n$ bins.

For example, for Lena image in Fig. 4.15, each R, G, B channel is represented by 8 bits and a total of 256 colors/bins. By concatenating the three histograms, the final histogram would have $N = 3 \times 256 = 768$ bins. This is, however, too long for image representation, therefore, the colors of each individual R, G, B channel is usually quantized to reduce the number of colors.

To quantize the color channels, colors in each channel are divided into equal intervals and each interval is used as a bin. For example, to create a 4-bin histogram for R channel, the 256 colors in R channel are divided into the 4 intervals: (0, 63), (64, 127), (128, 191), (191, 255). Figure 4.17 shows a 216-bin histogram by quantizing each of the R, G, B histograms in Fig. 4.15 into 72 bins. This is less than one-third of the length of the histogram without color quantization.

Fig. 4.16 RGB channels of a color image. Top row: R, G, B channels of the flower image; Bottom row: corresponding gray level images of the R, G, B channels at the top row

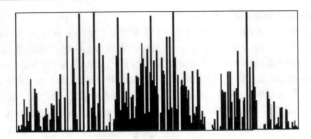

Fig. 4.17 Concatenation of histograms of individual R, G, B channels into a single histogram (216 bins, 72 bins for each channel)

4.4.1.2 Indexed Color Histogram

Another way to create a color histogram is to quantize the RGB color space (instead of each color plane) into N colors and use the N colors as bins to create a color histogram. This is equivalent to indexed colors and the N colors are equivalent to a global color palette, i.e., a palette representing all image colors or a palette for all the images in the world.

To quantize the RGB color space, the R, G, B planes are divided into l, m, n intervals, respectively, using the same way as in the first method, the RGB color space (a big cube) is then divided into $N = l \times m \times n$ small color cubes (Fig. 4.6). Each small cube represents a group of similar colors and is used as a histogram bin.

To create a histogram for a color image using the indexed colors, each pixel in the image is examined and put into a bin with similar colors to the pixel. A histogram is then created by counting the number of pixels in each of the bins. Figure 4.18a shows the flower image with 216 quantized or indexed colors and the 216-bin histogram for the quantized color image is shown in Fig. 4.18b.

4.4.1.3 Dominant Color Histogram

A histogram created from a global palette (a single fixed palette for all images) is usually sparse (Fig. 4.18b), this is because when a global palette is used, most of the colors in an image are often missing from the color representation, this effect is shown up in both the quantized image and the color histogram (Fig. 4.18a, b). In practice, an adaptive or native palette created from the image itself can be used, a histogram created from *adaptive indexed colors* is essentially a *dominant color histogram*. Figure 4.18c, d shows the flower image quantized with an adaptive palette of 216 colors and its histogram. Dominant colors can be obtained by either using a histogram thresholding or using a K-means clustering.

A histogram is invariant to translation and rotation changes, scale invariance can be achieved by normalizing each bin value with the total number of pixels in the image. The key issue with a histogram is the difficulty to determine the number of bins. If the number of bins is too small, the colors in a bin can vary so much that it causes too many confusions during the matching. On the other hand, if the number of bins is too large, it causes overfitting, means there are too few pixels in a bin, this too can cause confusions during the matching. In practice, there is always a

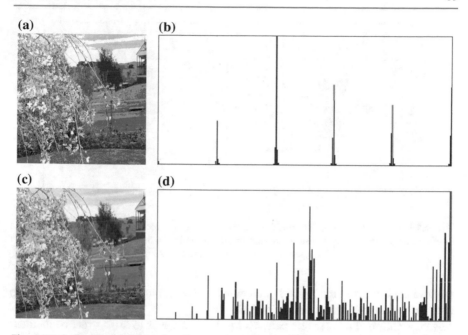

Fig. 4.18 Indexed color histograms of a color image. **a** The flower image quantized with a global palette of 216 indexed colors (notice the visible distortion on the sky and grass); **b** indexed color histogram of (**a**) (216 bins); **c** the flower image quantized with an adaptive palette of 216 indexed colors; **d** dominant color histogram of (**c**)

compromise on the number of bins. Regardless, features based on color histograms usually have very high dimensions, e.g., 512 and 1024 dimensions are common.

Another issue with the histogram method is that a color histogram does not tell pixels' spatial information. Therefore, visually different images can have similar color histograms. This is undesirable for image representation. A number of other color feature extraction methods have been designed to address these two issues, they are discussed in the next.

4.4.2   Color Structure Descriptor

One of the key drawback of a histogram is the absence of spatial information of pixel colors in an image. The spatial information of pixel colors tells the patterns of colors, or how colors are spread out inside an image. Without the spatial information of pixel colors, perceptually different images can have the same histogram and this can lead to incorrect image retrieval or classification. For example, Fig. 4.19 shows two perceptually different binary images with the same size and the same number of red pixels. The image on the left is visually structured and the red color is perceived as a regular shape, in contrast, the image on the right is visually

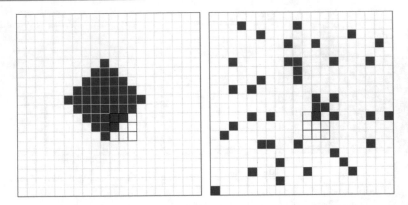

Fig. 4.19 Computation of color structure descriptor. Left: a color image with a 3 × 3 structure moving through the image. The structure captures the red color 56 times; Right: a color image with a 3 × 3 structure moving through the image. The structure captures the red color 218 times

unstructured and the red color is perceived as being cluttered. However, the two images have exactly the same color histogram.

One solution to detect color patterns inside an image is to use a structure element or window as the color picker instead of the pixel color picker/counter used in the ordinary histogram computation. When the window moves throughout an image, only the *colors* (e.g., white, red, gray, brown, etc.) inside the window are counted instead of counting the *pixels* of each color inside the window. The histogram created in this way is called color structure (CS) histogram. A CS histogram has a *multiplying* effect on the counting of isolated or scattered colors, the larger the structure window is used, the more the counting is multiplied. While the CS histogram only has a mild over-counting of grouped or clumped colors.

For example, in Fig. 4.19, both images have 41 red pixels, however, by moving a 3 × 3 structure window throughout the images, the red color in the right image are counted for 218 times, while the red color in the left image is only counted for 56 times. The large difference between the two figures accurately reflects the sharp difference between the two red patterns.

To create a CS histogram for an image, the image is first converted to HSV color space and is quantized into a smaller number of colors, e.g., 256, 128, 64, or 32 colors. A structuring element (e.g., square) is then moved throughout the image. Bin i of the histogram records how many times the structuring element captures at least one pixel with color i. If the window is of size 1 pixel, the CS histogram is just an ordinary histogram. In this sense, the ordinary histogram is just a special case of a CS histogram.

For example, the left-hand side of Fig. 4.20 shows a color image with 5 colors and the 4 × 4 window (black) capturing three types of colors: blue, green and brown [6]. The right-hand side shows how the CS histogram accumulates the three colors (green, blue. and brown) captured by the window into corresponding

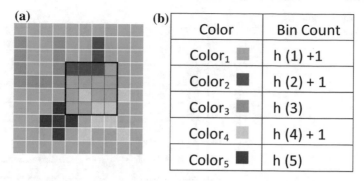

Fig. 4.20 Accumulation of color structure descriptor. **a** A 5-color image and a 4×4 structuring element. **b** The accumulation of color structure histogram at a particular position of the structuring element in the image

histogram bins (1, 2, and 4). The CS histogram is then normalized with a total number of counts of the histogram, and the normalized CS histogram is the CSD.

It can be expected that the CSD is more robust than an ordinary histogram, because it captures the information about local spatial structure as well as the color distribution of an image. The spatial information makes the CS histogram sensitive to certain color patterns to which an ordinary histogram is blind.

Furthermore, a CSD uses a window of size greater than 1 pixel, it is less susceptible to noise. However, the performance of CSD depends on the size and structure of the window. Scale invariance can only be achieved by varying the size of the structure element and doing the best match between two images. Rotation invariance can be achieved by using a circular element instead of a squared one.

4.4.3 Dominant Color Descriptor

It is understood that an image is visually interpreted based on a few dominant colors. Those other colors are either noise or just for details, they are not important and can be ignored. Therefore, a dominant color histogram will better describe an image than a common histogram. The *Dominant Color Descriptor* (DCD) is just based on this idea, it's a variation of a common histogram.

To derive a DCD, a histogram h of all colors (without quantization) in an input image I is first created. A thresholding is then applied to h to eliminate those bins whose values are less than a threshold τ. The remaining n colors are called dominant colors. Each dominant color is represented as $(\mathbf{c}_i, p_i)$, $i = 1, 2, \ldots, n$, where $\mathbf{c}_i$ is a 3D color vector and p_i is the percentage of pixels in the image having color $\mathbf{c}_i$. A DCD is just the n dominant colors in a sequence:

$$DCD = \{(\mathbf{c}_i, p_i), i = 1, 2, \ldots, n\} \tag{4.24}$$

Fig. 4.21 Statistics of DCD numbers from 36,692 regions

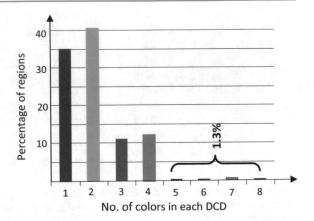

The DCD significantly reduces the dimensions of a color histogram, but it still does not address the absence of spatial information from the colors. Therefore, an image is usually segmented into regions and a DCD is extracted from each region of the image.

The number of selected dominant colors in a region depends on the threshold τ. However, statistics based on more than 36,000 image regions show that over 98% of image regions can be described by no more than 4 DCDs (Fig. 4.21) [6–8]. MPEG-7 recommends 1–8 DCDs for each image region.

Figure 4.22 shows some examples of the segmented image regions and their corresponding DCDs [6–8].

Small proportions of colors in a DCD are usually due to segmentation errors or region boundary, they can be discarded without affecting performance. Figure 4.23 shows a few segmented regions and their corresponding DCDs after discarding insignificant colors [6–8].

Region-based DCDs are not only compact but also reflect spatial information in an image, they are a desirable representation for color images. However, unlike conventional color histogram, the order of colors and the number of colors in two DCDs are usually not the same, the matching of two DCDs needs to use many-to-many quadratic matching (12.18).

DCDs can be easily translated into color names, which can be used to describe color images. Human beings tend to describe the visual world using color names, and we can only describe a few hundreds of colors. It is possible to annotate an image with color names based on DCDs [9].

4.4.4 Color Coherence Vector

An ordinary histogram does not tell spatial information about the colors in a bin. The *color coherence vector* (CCV) is a method to incorporate spatial information into a conventional histogram. The idea is to divide each histogram bin into two

Fig. 4.22 Segmented regions and their dominant colors underneath. The dominant colors are shown according to their percentages in the region

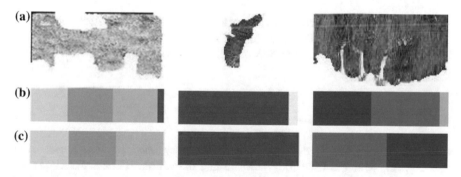

Fig. 4.23 Removal of noisy colors from segmented regions. **a** Three sample regions; **b** dominant colors of corresponding regions above; **c** DCDs after discarding insignificant colors from (**b**)

components: coherent (C) and noncoherent (N). The coherent component includes those pixels which are spatially connected, while the noncoherent component includes those pixels that are isolated. A CCV can be computed by using the following procedure:

1. Create a conventional histogram H of k bins for image I using a method in Sect. 4.4.1
2. For each of the histogram bins B_i in H

2.1. Create a binary image I_i from I by marking all pixels with color B_i as 1 (white) and others as 0 (black)

2.2. Set $j = 0$

2.3. For each white pixel p in I_i

2.3.1. If p's West, North West, North, and North East are all black

2.3.1.1. Create a new region R_j

2.3.1.2. $R_j = R_j + p$

2.3.1.3. $j = j + 1$

2.3.2. Otherwise, $R_j = R_j + p$

2.4. For each region R_j

2.4.1. Count the total number of pixels n in R_j

2.4.2. If $n \geq \tau$, $C_i = C_i + n$ //τ is a threshold

2.4.3. Otherwise, $N_i = N_i + n$

3. The normalized sequence $\{(\frac{C_i}{|I|}, \frac{N_i}{|I|}), i = 1, 2, \ldots, k\}$ is the CCV

By dividing each histogram bin into coherent colors and incoherent colors, CCV captures spatial information in an image, it usually performs better than a color histogram. However, the dimension of a CCV is twice of that of a conventional histogram.

4.4.5 Color Correlogram

A color *correlogram* is the color version of *gray level co-occurrence matrix* (Sect. 5.2.2), which is used for texture feature extraction. It characterizes the distribution of color pairs in an image. A color correlogram can be viewed as a 3D histogram, where the first two dimensions represent the colors of any pixel pair and the third dimension is their spatial distance. Thus, in a correlogram, each bin (i, j, k) represents the number of color pairs (i, j) at a distance k.

An input image I is first quantized into m colors $\{c_1, c_2, \ldots, c_m\}$. The computation of a color correlogram is then to find the probability of pixel pairs (p_1, p_2), which meet the following two conditions:

(a) $C(p_1) = c_i$ and $C(p_2) = c_j$

(b) $|p_1 - p_2| = k$

where $C(p_i) = c_i$ means the color of p_i is c_i.

Mathematically, each element of a correlogram is given by

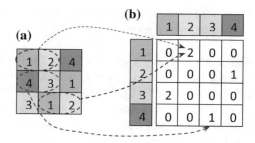

Fig. 4.24 Computation of color correlogram. **a** A 4-color image; **b** the color correlogram of **a** for horizontal distance $k = 1$

$$\gamma^k_{c_i,c_j}(I) = Pr\{(p_1,p_2)|C(p_1) = c_i, C(p_2) = c_j \, and \, |p_1 - p_2| = k\} \qquad (4.25)$$

If $c_i = c_j$, (4.25) becomes an autocorrelogram which is the probability of finding identical colors at distance k:

$$\alpha^k_c = r^k_{c,c}(I) \qquad (4.26)$$

Figure 4.24 shows an example of computing a color correlogram [6]. The color correlogram in Fig. 4.24 is calculated for $k = 1$. Correlograms for other distance $k \in \{1, 2, ..., d\}$ can be calculated in similar way.

In total, d correlograms (matrices) can be computed from an image. If the colors are globally quantized, i.e., colors of all images are quantized using a single global palette, two corresponding correlograms from two images can be matched element by element. However, if a local (adaptive) palette is used to quantize each image, either a many-to-many matching is needed or the matching is done through statistics computed from the correlogram matrices (Sect. 5.2.2).

Matching between two color correlograms involves matrix matching which is expensive. In practice, however, only autocorrelograms are used because they are sufficient to produce a good result. Autocorrelograms of each color form a d-dimensional vector, which can be matched using a conventional distance such as L_1 [10].

The performance of the color correlogram is better than the CCV, because it not only captures the special information but also the patterns of the spatial information.

4.4.6   Color Layout Descriptor

It is well known that a spectral transform can capture the frequency of texture changes in an image, and the frequency information is used for identifying most important information from the image. This idea can also be used to capture the frequency of color changes in an image. The *color layout descriptor* (CLD) is just based on this idea.

The computation of CLD consists of four stages: image partitioning, color quantization, DCT transform, and zigzag scanning.

- **Image partitioning**. In the first stage, an input image I is divided into an 8×8 grid of 64 blocks. If the size of the input image is $M \times N$, then the size of each block of the grid will be $(M/8) \times (N/8)$. The reason to divide all images into an equal number of blocks is to ensure resolution or scale invariance.
- **Color quantization**. In the second stage, a single dominant color is computed from each block. The DCD method in Sect. 4.4.2 can be used for the dominant color extraction, but the simplest method is to use the *average* of pixel colors as the representative color. Once the color of each block is quantized, the input image I is converted into an 8×8 color image I_q.
- **DCT transform**. In the third stage, the RGB colors of I_q are converted to Y'CbCr colors (4.21). Then, each of the three Y'CbCr channels of I_q is transformed by an 8×8 discrete cosine transform or DCT, so three sets of 64 DCT coefficients are obtained.
- **Zigzag scanning**. In the final stage, three sets of 64 DCT coefficients are zigzag scanned respectively and the first few coefficients of each set, e.g., 4–8, are chosen. The selected coefficients are then organized into (DY', DCb, DCr) which is used as the CLD. The reason of only choosing the first few coefficients is because they represent the low-frequency information of the image and they are the most significant coefficients, the remaining coefficients are too small and can be neglected.

CLD allows scalable representation of an image by controlling the number of selected coefficients. MPEG-7 recommends using a total of 12 coefficients, 6 for luminance and 3 for each chrominance, for most of the images. CLD is both compact and scalable, however, it is not robust to rotation change.

4.4.7   Scalable Color Descriptor

As can be seen from the CLD above, a spectral transform like DCT can dramatically reduce the data dimension. This idea can also be used to reduce histogram dimension. A histogram can be regarded as a 1D time series with fluctuations in the vertical direction. Each histogram has a unique pattern of fluctuations or changes along the horizontal direction. This pattern can be effectively captured by using efficient 1D wavelet transform. Because coefficients from a wavelet transform are scalable, i.e., the number of selected coefficients depends on requirement or applications, the result of the wavelet transform is a *scalable color descriptor* or SCD. A SCD is derived using the following procedure.

- **RGB to HSV**. An input image is first converted from RGB color space to HSV color space.

- **Color quantization**. The HSV color space is quantized into 256 colors by dividing the H, S, and V channels into 16, 4, and 4 intervals respectively. A 256-bin color histogram is then created for the image.
- **Bin value quantization**. Each bin value is then nonlinearly quantized into a 4-bit integer to give high significance to small values.
- **Harr wavelet transform**. The histogram is then applied with Harr wavelet transform. Each round of Harr wavelet is a two-pass transform, i.e., low pass and high pass. The low pass of Harr wavelet transform takes two neighboring bins and calculates their sum, while the high pass calculates their difference. Therefore, after the first round, two histograms with half of the original histogram length are obtained: a summed histogram and a differenced histogram. Repeat the transform on the summed histogram for a number of rounds until the two histograms are shortened to the desired length, e.g., 64, 32 or 16 bins.
- **SCD formation**. The final results from the wavelet transform are two short histograms: a summed histogram and a differenced histogram. The two histograms are concatenated to be used as the SCD. However, since the values of the differenced histogram bins are so small that the magnitudes of the bin values are discarded and only the signs of the bins are kept. The sign patterns are sufficient to retain the finer details of the original histogram.

SCD is useful for applications which need short or compact histogram features, however, it does not include spatial information as other color descriptors such as CLD, therefore, its performance is generally lower.

4.4.8   Color Moments

It is well known in data analysis community that descriptive statistics provide a good summary of a dataset and provide a quick understanding of the characteristics or distribution of the dataset such as central tendency, variability, skewness, etc. An image is just a set of color pixel data, therefore, it can also be described by the mean, variance, skewness, etc., which are called *color moments* (CM).

To compute color moments, an input image I is decomposed into individual channels, such as R, G, B channels or H, S, V channels. The moments are then computed from each channel using the following equations:

$$M_1 = \frac{1}{N} \sum_{i=1}^{N} p_i \tag{4.27}$$

$$M_r = \left(\frac{1}{N} \sum_{i=1}^{N} (p_i - M_1)^r \right)^{\frac{1}{r}} \tag{4.28}$$

where

- N is the total number of pixels in image I
- r is the order of a color moment, $r = 2, 3, \ldots$
- p_i is the ith pixel value in the color channel
- M_1 is the first-order color moment, or the mean color of the color channel
- M_2 is the second-order color moment, or the variance of the color channel
- M_3 is the third-order color moment, or the skewness of the color channel

Color moments can also be computed from a color histogram h using the following equations:

$$M_1 = \sum_{k=1}^{K} h_k C_k \qquad (4.29)$$

$$M_r = \left(\sum_{k=1}^{K} h_k (C_k - M_1)^r \right)^{\frac{1}{r}} \qquad (4.30)$$

where

- h_k is the value of the kth bin of histogram h
- C_k is the color of kth bin of histogram h
- K is the number of bins of h_q
- r is the order of a color moment: $2, 3, \ldots$

Typically, only the first three order color moments are computed for a color channel or an image. If three color moments are computed for each color channel, the moments from each of the three channels are concatenated to form a 9-dimensional feature vector which is used to describe the image.

Color moments are a very concise description of an image, however, it can be very inaccurate, e.g., the mean or average color of an image is usually a very coarse description of the image color. Furthermore, color moments do not tell the spatial information of the colors. Therefore, color moments are usually calculated for image regions.

4.5   Summary

Color is often the first feature to be considered during image processing and analysis. A number of preprocessing or preparation are usually performed before the actual color feature extraction such as color space conversion, noise reduction, image scaling, clustering, and segmentation.

A good understanding and choice of color space can make a significant difference in the performance of the extracted color features. Usually, HSV, LUV, and Y'CbCr color spaces are preferred to RGB color space. This is because unlike RGB color space, channels of HSV, LUV, and Y'CbCr color spaces are independent of each other. The independence of color channels leads to less confusions in image classification and retrieval.

Color feature extraction typically starts with color quantization, which aims to remove insignificant colors from an image and increase the robustness of the extracted color features.

Histogram plays a key role in color feature extraction, most of the color features are histogram based or histogram related. These methods usually aim at either reducing the dimensions of a histogram, e.g., SCD, CM; or incorporating spatial information into a histogram, e.g., CCV, CSD, correlogram; or both, e.g., DCD.

A histogram itself is a scalable descriptor, because as shown in SCD, neighboring bins can be merged to shorten a histogram to a desirable length. Histograms are usually invariant to rotation and translation, they can also be normalized to scale invariance using the total number of pixels in an image or a region.

Histograms created from color channels and indexed colors can be matched using a simple distance measure such as L_p. However, histograms created from dominant colors need to be matched using a many-to-many matching such as quadratic distance. This is because colors of corresponding bins of two dominant color histograms are different.

The computation cost of color descriptors ranges from low such as histogram, CM, to moderate such as DCD, SCD, CSD, to high such as correlograms, CCV, CLD.

4.6 Exercises

1. Find a gray level image, use the Matlab code from the following webpage to generate the histogram of the image: https://au.mathworks.com/help/images/ref/imhist.html. Explain the values on both the horizontal and vertical axes.
2. Normalize the above histogram values to between 0 and 1 to convert the histogram into a feature vector.
3. Apply logarithm transform to the histogram in problem 1, compare the log-transformed histogram with the original histogram. Explain the similarity and difference between the two histograms, tell how useful is the log-transformed histogram in image data mining.
4. Generate histograms for similar images with different brightness, images with different patterns, tell the pros and cons of using image histograms as image features.
5. Find a color image, open GIMP image processing software (free software from gimp.org), use command: Colors → Components → Decompose (uncheck "Decompose to layers") to decompose the color image into RGB, HSV, Lab and

YCbCr channels. Write a short report to contrast the difference between the 4 color spaces and tell the applications of the 4 color spaces.
6. Google online using "JSEG (Matlab implementation) Y. Deng" and download the Matlab code of JSEG from the Software's site http://cs.joensuu.fi/~zhao/Software/, find a color image and run the "script.m" file in Matlab editor (change the path to your image file such as filename = ["..\images\flower.jpg"]). Analyze the segmentation results. Try more images and explain how well it can be used for color feature extraction.

References

1. Wikipedia (2019) Young–Helmholtz theory. https://en.wikipedia.org/wiki/Young%E2%80%93Helmholtz_theory. Accessed Feb 2019
2. Stanford University (2019) EE386 lectures. https://web.stanford.edu/class/ee368/Handouts/Lectures/Examples/. Accessed Feb 2019
3. Abraham C (2019) A beginner's guide to (CIE) colorimetry. https://medium.com/hipster-color-science/a-beginners-guide-to-colorimetry-401f1830b65a. Accessed Feb 2019
4. Stanford University (2018) EE386 lectures. https://web.stanford.edu/class/ee368/Handouts/Lectures/2018_Winter/13-ScaleSpace.pdf. Accessed Oct 2018
5. Deng Y, Manjunath BS, Shin H (1999) Color image segmentation. In: Proceedings of CVPR'99, vol 2, pp 446–451
6. Islam M (2009) SIRBOT—semantic image retrieval based on object translation. PhD thesis, Monash University
7. Zhang D, Islam M, Lu G (2013) Structural image retrieval using automatic image annotation and region based inverted file. J Vis Commun Image Represent 24(7):1087–1098
8. Islam M, Zhang D, Lu G (2008) Automatic categorization of image regions using dominant color based vector quantization. In: Proceedings of digital image computing: techniques and applications (DICTA 2008), pp 191–198
9. Liu Y, Zhang D, Lu G, Ma WY (2005) Region-based image retrieval with high-level semantic color names. In: Proceedings of multimedia modelling conference, pp 180–187
10. Huang et al (1997) Image indexing using color correlograms. In: Proceedings of CVPR 97

Texture Feature Extraction

5

The devil is in the detail.

5.1 Introduction

Texture is a general pattern that can be attributed to almost everything in nature. For a human, texture patterns relate to specific and spatially repetitive structure of surfaces formed by repeating a particular element or several elements in different spatial positions. Generally, the repetition involves local variations of scale, orientation, or other geometric and optical features of the elements.

Texture is an inherent feature of an object. For example, we can easily tell if an object surface is fine or rough, regular or natural, quiet or busy, etc. It is found that human beings tend to recognize texture by its structure or how often it changes. As the result, the texture methods designed in the last few decades are along two directions: spectral methods and spatial methods. Spatial texture methods attempt to capture the primitive patterns of objects and compute the structural features; while spectral texture methods attempt to capture the change patterns of objects and compute the frequency of changes. Spatial methods are generally more intuitive, while spectral methods are generally more efficient and robust. In this chapter, we discuss those important texture methods for image representation.

5.2 Spatial Texture Feature Extraction Methods

In spatial approach, texture features are extracted by computing the pixel statistics or finding the local pixel structures in the original image. These methods include the Tamura textures, co-occurrence matrix method, Markov random field (MRF) method, and fractal dimension (FD) method. Tamura et al. are among the earliest researchers to formally define texture features [1]. The most cited Tamura texture features in literature consist of six perceptual characteristics of images such as the degree of contrast, coarseness, directionality, linearity, roughness, and

© Springer Nature Switzerland AG 2019
D. Zhang, *Fundamentals of Image Data Mining*, Texts in Computer Science, https://doi.org/10.1007/978-3-030-17989-2_5

regularity. In most of the cases, only the first three Tamura features are used as the other three features are defined based on the combinations of the first three features. Tamura features are nice because they are high-level perceptual features and suitable for texture browsing. However, it is difficult to define more such types of high-level features. Therefore, Tamura features are not enough to distinguish all the textures in the world.

5.2.1 Tamura Textures

Tamura et al. [1] introduce six statistical features. These include *coarseness, contrast, directionality, line-likeness, regularity*, and *roughness*. The last three features are defined based on the first three features. Therefore, most of the image retrieval systems only use the first three Tamura features.

Coarseness relates to the size of the primitive elements (textons) forming the texture, and it measures the image granularity. It is calculated as the average of the largest window sizes needed to identify texture elements centered at different pixel positions. Formally, it is defined as

$$f_{crs} = \frac{1}{n^2} \sum_{x=1}^{n} \sum_{y=1}^{n} 2^k I(x, y) \tag{5.1}$$

where $n \times n$ denotes the image size of $I(x, y)$, and k is obtained as the value which maximizes the differences of the moving averages of $A_k = \frac{1}{2^{2k}} \sum_{x=1}^{n} \sum_{y=1}^{n} I(x, y)$, taken over a $2^k \times 2^k$ neighborhood along the horizontal and vertical directions. The specific procedure is to compute the differences between the average signals for the nonoverlapping windows of different size:

(1) At each pixel (x, y), compute six averages for the windows of size $2^k \times 2^k$, $k = 0, 1, \ldots, 5$, around the pixel.
(2) At each pixel, compute absolute differences $E_k (x, y)$ between the pairs of nonoverlapping averages A_k in the horizontal and vertical directions.
(3) At each pixel, find the value of k that maximizes the difference $E_k (x, y)$ in either direction and set the best size $S_{best} (x, y) = 2^k$.
(4) Compute the coarseness feature f_{crs} by averaging $S_{best} (x, y)$ over the entire image. Textures with multiple coarseness can be computed from the histogram of $S_{best} (x, y)$.

Contrast tells how well an object is distinguishable from other objects or background. It measures how gray levels q vary in the image I and to what extent their distribution is biased to black or white. The second-order σ^2 and normalized fourth-order central moments μ_4 of the gray level histogram (empirical probability distribution P) are used to define the contrast feature:

$$f_{con} = \frac{\sigma}{\left(\frac{\mu_4}{\sigma^4}\right)^{\frac{1}{4}}} \tag{5.2}$$

where $\mu_4 = \sum_{q=0}^{q_{max}} (q - m)^4 P(q|I)$ is the *kurtosis*; $\sigma^2 = \sum_{q=0}^{q_{max}} (q - m)^2 P(q|I)$ is the *variance*, and m is the *mean* gray level.

Directionality tells if there exists any directional pattern in an image, like vertical, horizontal, diagonal, etc. The degree of directionality is measured using the frequency distribution of oriented local edges against their directional angles. The edge strength $e(x, y)$ and the directional angle $a(x, y)$ are computed to approximate the pixelwise x and y derivatives of the image:

$$e(x, y) = \left(|\Delta_x(x, y)| + |\Delta_y(x, y)|\right)/2 \tag{5.3}$$

$$\phi(x, y) = arctan(\Delta_y(x, y)/\Delta_x(x, y)) \tag{5.4}$$

where $\Delta_x(x, y)$ and $\Delta_y(x, y)$ are the horizontal and vertical gray level differences between the neighboring pixels, respectively. They are computed by using Prewitt edge detectors

Δ_x			Δ_y		
−1	0	1	1	1	1
−1	0	1	0	0	0
−1	0	1	−1	−1	−1

A histogram $h_{dir}(\phi)$ of quantized direction values ϕ is constructed by counting the numbers of the edge pixels with the corresponding directional angles and the edge strength greater than a predefined threshold. The histogram is relatively uniform for images without strong orientation and exhibits peaks for highly directional images. The directionality feature is defined as the *sharpness* of the histogram:

$$f_{dir} = 1 - rn_p \sum_{p=1}^{n_p} \sum_{\phi \in w_p} (\phi - \phi_p)^2 h_{dir}(\phi) \tag{5.5}$$

where n_p is the number of peaks, ϕ_p is the position of the pth peak, w_p is the range of the angles attributed to the pth peak (that is, the range between valleys around the peak), r denotes a normalizing factor related to quantising levels of the angles ϕ, and ϕ is the quantized directional angle.

Figure 5.1b shows the edge map of an original image in Fig. 5.1a. To compute f_{dir}, h_{dir} is first computed by quantizing ϕ and counting the number of edge pixels with $e(x, y)$ greater than a threshold and the edge histogram of angles is shown in Fig. 5.1c. The angles ϕ in the horizontal axis are in the range of −90° to +90° and

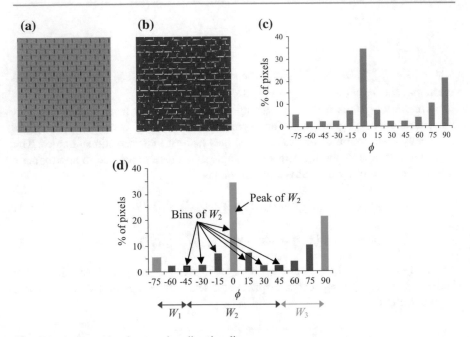

Fig. 5.1 An example of computing directionality

are quantized into 12 intervals and the quantized angles are $-75°$, $-60°$, $-45°$, …, $+90°$. The vertical axis shows the percentage of edge pixels at different angles [2].

After computing the h_{dir}, all peaks and valleys in h_{dir} are detected. Figure 5.1d shows the peaks and valleys in green and blue, respectively. Suppose, there are n_p peaks in the histogram. For each peak p, let w_p be the window of bins from the previous valley to the next valley (a window contains a peak in it), and ϕ_p be the angular position of the peak in w_p. Based on the definition of f_{dir}, the more directional an image, the higher the directionality the image. However, f_{dir} is not invariant to rotation, rotation of an image causes a circular shift of the h_{dir} histogram, and this can cause false peak detection. For example, the first peak in Fig. 5.1d is actually a part of the hill defined by the last peak of the histogram. Therefore, in practice, several rounds of circular shift are needed to find out the real peaks of the histogram.

Tamura textures are intuitive in terms of definitions, however, the computation processes are complex. This affects the robustness and the overall performance of the computed features.

5.2.2   Gray Level Co-occurrence Matrices

Many statistical texture features are based on gray level co-occurrence matrices (GLCM) or its color counterpart color correlogram. A GLCM represents how frequent is every particular pair of gray levels in an image, separated by a certain distance d along a certain direction a.

Formally, given an $n \times m$ image $I(x, y)$, a cell in a GLCM is defined as

$$C_{\Delta x,\Delta y}(i,j) = \sum_{x=1}^{n} \sum_{y=1}^{m} \begin{cases} 1, & \text{if } I(x,y) = i \text{ and } I(x+\Delta x, y+\Delta y) = j \\ 0, & \text{otherwise} \end{cases} \tag{5.6}$$

For an image with 256 gray level values, a GLCM is a 256×256 matrix. With the combination of Δx and Δy, a large number of GLCMs can be created. In practice, only four GLCMs are created by capturing the following four structures: horizontal, vertical, left lean diagonal, and right lean diagonal.

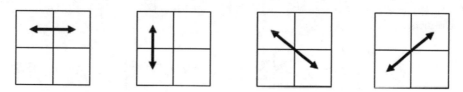

Figure 5.2 shows an example GLCM of an image $I(x, y)$ with 8 gray level values. In this case, the structure to be captured by the GLCM is the *horizontal* structure. The arrows in the figure link the pixel pairs in the image with their corresponding entries in the GLCM.

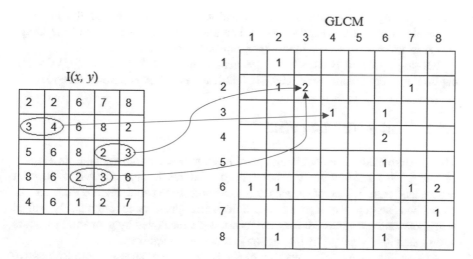

Fig. 5.2 Computation of a GLCM. An 8 gray levels image $I(x, y)$ on the left and its GLCM on the right

A GLCM itself is a gray level image, therefore, a number of statistical features called Haralick features can be computed from each GLCM image, such as the homogeneity, contrast, correlation, energy, entropy, variance, etc. [3].

Suppose $p(i, j)$ is the normalized value of a GLCM entry, it is equivalent to the probability of a particular structural distribution in the image; N_g is the number of gray levels in the quantized image. The three major texture features *angular second moment*, *contrast*, and *correlation* are given in the following:

Angular second moment:

$$f_1 = \sum_i \sum_j \{p(i,j)\}^2 \qquad (5.7)$$

Contrast:

$$f_2 = \sum_{n=0}^{N_g-1} n^2 \left\{ \sum_i \sum_j [p(i,j) | |i - j| = n] \right\} \qquad (5.8)$$

Correlation:

$$f_3 = \frac{\sum_i \sum_j (ij) p(i,j) - \mu_x \mu_y}{\sigma_x \sigma_y} \qquad (5.9)$$

where μ_x, μ_y, σ_x, and σ_y are the means and standard deviations of the marginal probabilities p_x and p_y, respectively. f_1 is a measure of homogeneity of the image and f_3 is a measure of gray tone linear dependencies in the image.

Because a GLCM is usually a large matrix and needs a number of GLCMs to capture different texture structures, GLCM features are expensive to compute.

5.2.3 Markov Random Field

MRF texture methods model image pixel location as a random variable, as a result, an image is a random field. Each type of textures is characterized by a joint probability distribution of signals that accounts for spatial interdependence, or interaction among the signals. The interacting pixel pairs are usually called neighbors, and a random field texture model is characterized by geometric structure and quantitative strength of interactions among the neighbors.

Among the many MRF texture methods, the Simultaneous Auto-Regressive (SAR) model is the most widely used, as it uses fewer parameters. In SAR, the intensity $I(x, y)$ at pixel (x, y) is estimated as a linear combination of the neighboring pixel values $I(s, t)$ and an additive Gaussian noise $\varepsilon(x, y)$.

$$I(x,y) = \mu + \sum_{(s,t) \in N} \theta(s,t)I(s,t) + \varepsilon(x,y) \tag{5.10}$$

where

- μ is the mean of the image,
- N is the neighborhood of (x, y), e.g., a 3×3 window,
- $\theta(s, t)$ are the weights or coefficients associated with each of the neighborhood pixels,
- and $\varepsilon(x, y)$ is a Gaussian error with zero mean and standard deviation of σ.

The set of parameters θ and σ are the measurement of the texture, they can be estimated using either the least square error (LSE) technique or the Maximum Likelihood Estimation (MLE). Both LSE and MLE involve complex optimization which is computationally expensive. A higher σ value indicates finer granularity or less coarseness; a higher $\theta(x, y + 1)$ and $\theta(x, y - 1)$ values indicate that the texture is vertically oriented, so on so forth.

Rotation-Invariant SAR model (RISAR) can be created by replacing N with a circular neighborhood. In order to make SAR more robust, Multiresolution MRF (MRMRF) can also be created, where an image is represented by a multiresolution Gaussian pyramid before applying the MRF model.

The number of parameters or the feature dimensions of a SAR depends on the size of the neighborhood, e.g., a 3×3 neighborhood results in 9 parameters while a 4×4 neighborhood results in 16 parameters. To be scale invariant, SARs with multiple window size are needed, this makes the computation of MRF features prohibitively expensive.

5.2.4   Fractal Dimension

The fractal dimension (FD) method [4] is based on the theory of fractal geometry which characterizes the shapes or patterns of self-similarity. The idea of fractal is to find the smallest structure which replicates the whole pattern. According to fractal theory, a bounded set S in Euclidean space R^n is self-similar if S is the union of N (r) distinct (nonoverlapping) copies of itself scaled up or down by a ratio r, and the relationship between $N(r)$ and r is given by

$$N(r) \approx C \left(\frac{1}{r} \right)^d \tag{5.11}$$

where d is the fractal dimension or FD.

In image applications, FD method models a gray level image as a 3D terrain surface, and a differential box counting is done under the surface to measure how rough the surface is. In logarithm term, the above relationship means the number of boxes under the surface is inversely proportional to the size of the boxes, which is expected. d is given by the following approximation:

$$d = \lim_{r \to 0} \frac{\log N(r)}{\log \frac{1}{r}} \tag{5.12}$$

or

$$d \approx \frac{\log N(r)}{\log \frac{1}{r}} \tag{5.13}$$

From (5.13), the $\log N(r)$ and $\log(1/r)$ have an approximately *linear* relationship, therefore, FD can be estimated from a least square fitting of the two variables. By fitting a straight line for the $\log N(r)$ versus $\log(1/r)$ curve, the *slope* of the straight line is taken as the approximation of FD.

Since FD only models the roughness feature, other features like directionality and contrast are missed from FD. Therefore, in [4], six FDs have to be computed from a number of modified images derived from the original image, such as, the original image, low gray-valued image, high gray-valued image, horizontally smoothed image, vertically smoothed image, and the second moment of the original image. Despite of these additional FD features, FD can be very sensitive due to the triple approximation during the box counting, linear fitting, and image modification.

5.2.5 Discussions

Spatial texture methods are based on the ideas of capturing the elemental or microstructures of a textured image. The definitions of the structures are based on how humans describe a textured image, such as rough versus fine, regular versus natural, directional versus random, etc. The advantage of these methods is that they are intuitive and semantically meaningful. However, there are infinite types of textural structures in the nature, and human beings can only define or describe a small number of them. This can limit the application of spatial texture methods.

Another major issue with spatial approach is that spatial features are sensitive to noise. Furthermore, spatial texture methods are usually complex to compute, and they often involve complex optimization which is very expensive to compute. These issues affect the robustness and the overall performance of spatial texture methods.

5.3 Spectral Texture Feature Extraction Methods

Instead of defining and describing specific structures in an image, which is difficult, spectral texture methods attempt to capture how frequent the patterns change in a textured image. Spectral texture methods are based on Fourier Transform (FT), Discrete Cosine Transform (DCT), Discrete Wavelet Transform (DWT), Gabor filters, curvelet transform, etc.

Global power spectra computed from the DFT are not effective in texture classification and retrieval, compared with local features computed from small windows such as DCT. At present, the most promising features for texture retrieval are based on multiresolution features obtained with orthogonal wavelet transforms or Gabor filters. These features describe spatial distributions of oriented edges in the image at multiple scales.

5.3.1 DCT-Based Texture Feature

Compared with the traditional spatial texture methods, DCT is a simple yet robust method to capture local textures of an image. The idea is equivalent to STFT or applying FT on a small window. However, due to the use of 1D cosine transform on both rows and columns, the computation is very efficient.

For a color image I, it is first converted to Y'CbCr or YBR colors. The image is then divided into a set of overlapping 8×8 regions or blocks, which are obtained by a sliding window that moves by two pixels between consecutive samples. At each location of the three YBR color channels, apply the DCT on the local 8×8 window. Each block is then represented by

$$\mathbf{x} = \left[\mathbf{x}^Y, \mathbf{x}^B, \mathbf{x}^R\right] \tag{5.14}$$

where $[\mathbf{x}^Y, \mathbf{x}^B, \mathbf{x}^R]$ is the concatenation of the DCT vectors extracted from each of the YBR color channels by a zigzag scanning. For efficient computation, the 192-dimensional YBR-DCT vector is usually shortened by only retaining the first few coefficients from each of the YBR channels. This is because of the well-known energy compaction properties of the DCT.

To compute the texture features of the image I, a Gaussian mixture model of eight components is computed using the EM algorithm. This produces the following conditional distribution for each image:

$$P(\mathbf{x}|I) = \sum_{k=1}^{8} \pi_I^k G(\mathbf{x}, \mu_I^k, \sigma_I^k) \tag{5.15}$$

where π_I^k is the weight and μ_I^k and σ_I^k are the maximum likelihood parameters of mixture component k. The (μ_I^k, σ_I^k) pair is then organized into a feature vector which is used for texture representation. For a gray level image, the DCT method just needs to replace the three color channels with a single gray channel.

DCT computation is efficient due to the use of FFT, however, the EM is an optimization method which incurs significant computation cost.

5.3.2 Texture Features Based on Gabor Filters

5.3.2.1 Gabor Filters

Although DCT is efficient to compute locally, the image level texture features are complex to compute due to the use of EM algorithm. An alternative is to use Gabor filters. Gabor filters are based on traditional filter-based image processing approach, which computes one filtered value at each pixel as opposing to computing multiple transformed values at each location as in the DCT. Different from traditional filters, by combining both Gaussian and FT, Gabor filters simulate the powerful properties of perceptual vision of mammals. Furthermore, they can be tuned to different orientations and scales.

Gabor transform creates a filter bank consisting of Gabor filters with various scales and orientations. For a given image $I(x, y)$ with size $P \times Q$, its discrete Gabor transform is given by a convolution:

$$G_{mn}(x, y) = \sum_{s=0}^{K} \sum_{t=0}^{K} I(x - s, y - t) g_{mn}^*(s, t) \tag{5.16}$$

where K is the filter mask size, and g_{mn}^* is the complex conjugate of g_{mn} which is a class of self-similar wavelets generated from dilation and rotation of the following mother wavelet:

$$g(x, y) = \frac{1}{2\pi\sigma_x\sigma_y} \exp\left[-\frac{1}{2}\left(\frac{x^2}{\sigma_x^2} + \frac{y^2}{\sigma_y^2}\right)\right] \cdot \exp(j2\pi Wx) \tag{5.17}$$

where W is called the modulation frequency. The self-similar Gabor wavelets are obtained through the generating function

$$g_{mn}(x, y) = a^{-m} g(\tilde{x}, \tilde{y}) \tag{5.18}$$

where m and n specify the *scale* and *orientation* of the wavelet respectively, with $m = 0, 1, \ldots M - 1$, $n = 0, 1, \ldots, N - 1$, and

$$\left. \begin{aligned} \tilde{x} &= a^{-m}(x\cos\theta + y\sin\theta) \\ \tilde{y} &= a^{-m}(-x\sin\theta + y\cos\theta) \end{aligned} \right\} \tag{5.19}$$

where $a > 1$ and $\theta = n\pi/N$.

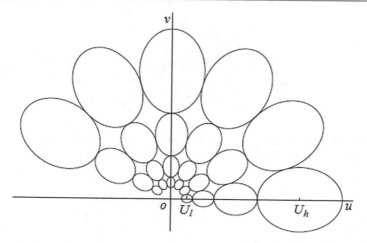

Fig. 5.3 FWHM sampling of spectral responses of Gabor filters in frequency plane

In order to decide the bank of Gabor filters, the parameters in (5.17) to (5.19) have to be determined. The Gabor wavelets generated using (5.17) through (5.19) are complete but not orthogonal wavelets, it implies there is redundancy in the filters. Therefore, we may design a bank of Gabor filters so that redundancy will be significantly reduced while image texture will be well represented by the set of individual filter response. Due to the same functional form of Gabor wavelet function $g(x, y)$ and its frequency response $\Psi(u, v)$ (i.e., its 2-D Fourier transform):

$$\Psi(u, v) = \exp\{-\frac{1}{2}[\frac{(u - W)^2}{\sigma_u^2} + \frac{v^2}{\sigma_v^2}]\} \tag{5.20}$$

where $\sigma_u = (2\pi\sigma_x)^{-1}$ and $\sigma_v = (2\pi\sigma_y)^{-1}$, an optimal representation of an image in spatial domain can be achieved by finding an optimal representation of the image in frequency domain. The idea of achieving so is to use the full width at half maximum (FWHM) of the Gabor spectral functions to form a complete coverage of the frequency plane within the modulation frequency bandwidth (Sect. 2.2). Specifically, the design follows three principles [5]: (i) *Uniform sampling* of orientation angles; (ii) *Exponential sampling* of modulation bandwidth W; (iii) Continuous coverage of the frequency space. This strategy is illustrated in Fig. 5.3.

Let U_l and U_h be the lowest and highest frequencies of interest, such that the coarsest scale filter and the finest scale filter are centered in the frequency domain at distances U_l and U_h from the origin, respectively. By the above strategy of redundancy reduction, the parameters of Gabor filters in spatial domain can be determined as follows:

$$a = (U_h/U_l)^{\frac{1}{M-1}} \tag{5.21}$$

$$\sigma_{x,m,n} = \frac{(a+1)\sqrt{2\ln 2}}{2\pi a^m(a-1)U_l} \tag{5.22}$$

$$\sigma_{y,m,n} = \frac{1}{2\pi \tan\left(\frac{\pi}{2N}\right)\sqrt{\frac{U_h^2}{2\ln 2} - \left(\frac{1}{2\pi\sigma_{x,m,n}}\right)^2}} \tag{5.23}$$

The parameters are independent of orientations (n), in other words, the parameters repeat in every orientation. In practice, the following parameter values are used:

$$U_l = 0.05, U_h = 0.4, K = 60$$

5.3.2.2 Gabor Spectrum

Figure 5.4 shows the Gabor filtered subband images from the lady image of Fig. 3.2. It shows how different image features are captured by Gabor filters from different scales and orientations. It can be observed that low-frequency information are captured at lower scales and as scale increases (*in spectral domain*), more fine details can be seen.

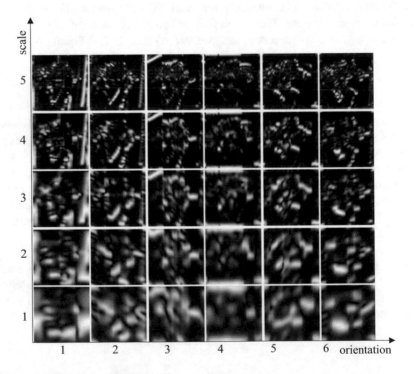

Fig. 5.4 Gabor filtered subbands for the lady image

5.3.2.3 Texture Representation

After applying Gabor filters on the image with different orientation at a different scale, a set of magnitudes is obtained:

$$E(m,n) = \sum_{x=0}^{P} \sum_{y=0}^{Q} |G_{mn}(x,y)|, m = 0, 1, \ldots, M-1; \; n = 0, 1, \ldots, N-1 \quad (5.24)$$

where

- m is the scale
- n is the orientation
- M is the maximal scale
- N is the maximal orientation
- $P \times Q$ is the size of the input image

The magnitudes represent the energy map at a different scale and orientation of the image under transform (Fig. 5.5 right) [6].

The main purpose of texture-based retrieval is to find images or regions with similar texture. It is assumed that we are interested in images or regions that have homogenous texture, therefore the following mean μ_{mn} and standard deviation σ_{mn} of the magnitude of the transformed coefficients are used to represent the homogenous texture feature of the image or region:

$$\mu_{mn} = \frac{E(m,n)}{P \times Q}$$

$$\sigma_{mn} = \frac{\sqrt{\sum_x \sum_y \left(|G_{mn}(x,y)| - \mu_{mn}\right)^2}}{P \times Q} \quad (5.25)$$

A feature vector $\mathbf{f}$ (texture representation) is created using μ_{mn} and σ_{mn} as the feature components. Five scales and six orientations are used in common implementation and the Gabor texture feature vector is thus given by

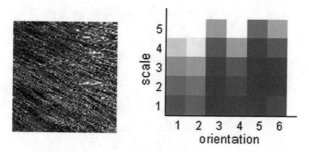

Fig. 5.5 Computation of Gabor texture descriptor. A straw image on the left and its energy map on the right. The higher the energy the brighter the block

$$\mathbf{f} = (\mu_{00}, \sigma_{00}, \mu_{01}, \sigma_{01}, \ldots, \mu_{45}, \sigma_{45}) \tag{5.26}$$

In order to remove the influence of various lighting issues of the camera, the features μ_{mn} and σ_{mn} can be normalized to [0, 1] using the maximum of the respective components. The similarity between two texture patterns is measured by the city block or Euclidean distance between their Gabor feature vectors.

5.3.2.4   Rotation-Invariant Gabor Features

The above-acquired texture feature is not invariant to rotation, similar texture images with different direction may be missed out from the retrieval or get a low rank. A simple circular shift on the feature map can be used to solve the rotation variant problem. Specifically, the orientation with the highest energy is detected as the dominant direction of the image and the feature elements in the dominant direction are moved to the first elements in $\mathbf{f}$. The other elements are then circularly shifted accordingly. For example, if the original feature vector is "*abcdef*" and "*c*" is at the dominant direction, then the normalized feature vector will be "*cdefab*" [7].

This circular shift approach is based on the theory that image rotation in spatial domain is equivalent to circular shift in spectral domain. Assume the original image is $I(x, y)$, $I_\phi(x, y)$ is the result of $I(x, y)$ after rotation of angle ϕ, by using (1.40), we have the following:

$$I_\phi(x, y) = I(x, y) \cdot e^{-j\phi} \tag{5.27}$$

For notation convenience, the Gabor transform of $I(x, y)$ and $I_\phi(x, y)$ at scale s and angle θ are denoted as $G(I, s, \theta)$ and $G(I_\phi, s, \theta)$, respectively. Then according to (5.16), we have

$$G(I, s, \theta) = I(x, y) * g_{s\theta}(x, y) \tag{5.28}$$

and by the commutability of convolution, we have

$$\begin{aligned} G(I_\phi, s, \theta) &= I_\phi(x, y) * g_{s\theta}(x, y) \\ &= \left[I(x, y) \cdot e^{-j\phi} \right] * g_{s\theta}(x, y) \\ &= I(x, y) * \left[g_{s\theta}(x, y) \cdot e^{-j\phi} \right] \end{aligned} \tag{5.29}$$

Equation (5.29) indicates that applying a Gabor filter on a rotated image is equivalent to applying a rotated Gabor filter on the original image. Since $g_{s\theta}(x, y) \cdot e^{-j\phi} = g_{s,\theta+\phi}(x, y)$, we have

$$\begin{aligned} G(I_\phi, s, \theta) &= I(x, y) * \left[g_{s\theta}(x, y) e^{-j\phi} \right] \\ &= I(x, y) * g_{s,\theta+\phi}(x, y) \\ &= G(I, S, \theta + \phi) \end{aligned} \tag{5.30}$$

Fig. 5.6 Computation of
rotation-invariant Gabor
texture descriptor. **a** A straw
image; **b** energy map of (**a**);
c a rotated image of (**a**);
d energy map of (**c**). The
higher the energy, the brighter
the block

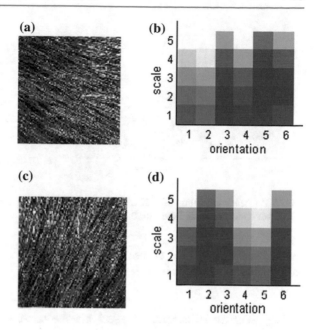

Equation (5.30) indicates that a rotation of the input image $I(x, y)$ by an angle ϕ
is equivalent to a translation of the output energy $G(I, s, \theta)$ by the same amount ϕ
along the orientation axis. Figure 5.6 demonstrates this fact. It shows two texture
patterns and their feature maps, pattern (c) is a rotation of 90° of pattern (a). Form
feature map (b), it can be seen that pattern (a) has a dominant direction feature in
orientation 2 (60°), while in feature map (d), this dominant direction feature has
moved to orientation 5 (150°) and features in other directions are circularly shifted
accordingly. In other words, the spectrum (d) is the circularly shifted version of
spectrum (b) [6].

If a texture pattern has directional features, it will show dominant energy at
certain direction on the energy map. If the direction of the highest energy is circularly shifted to zero degree, the resulting **f** is a rotation-invariant feature. If a
texture pattern does not have dominant direction feature, the matching between
rotated patterns can be made at any direction, and the rotation normalization does
not affect the matching in this situation.

In image analysis, there is always a compromise between spatial resolution and
frequency resolution. Gabor filters achieve optimal joint localization/resolution in
both space domain and frequency domain. However, due to the truncation at half
peak magnitude, the spectral cover of Gabor filters is not complete, this results in
information loss in the spectral domain. For example, in Fig. 5.7, black holes are
left at the FWHM in Gabor transformed spectral domain. Consequently, the
high-frequency components, which are considered to be the most important in
characterizing image textures, are not completely captured. Abundant redundancy
also exists between transformed images at different scales because Gabor filters use

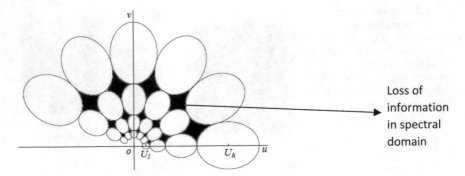

Fig. 5.7 Frequency tiling of half frequency plan by Gabor filters, the ovals are the covered spectrum while the black holes are the lost spectrum

overlapped window and do not involve image down sampling. These limitations can be easily overcome by using wavelet transform.

5.3.3   Texture Features Based on Wavelet Transform

5.3.3.1   Selection and Application of Wavelets

The key idea of wavelet transform is to analyze an image using multiresolution approach by using filters of different size, called wavelets. This is equivalent to look at the image from different distance. The whole decomposition process provides us with an array of DWT coefficients obtained from each subbands at each scale. These coefficients can then be used to represent the texture features of an image.

Given a 2D image $f(m, n)$, $0 \leq m \leq M - 1$, $0 \leq n \leq N - 1$, its DWT is given by (5.31):

$$W^j(k, l) = \frac{1}{2^j} \sum_{n=0}^{N-1} \sum_{m=0}^{M-1} f(m, n) \psi_{k,l}^j(m, n) \tag{5.31}$$

where j is the scale and (k, l) is the spatial location of the wavelet and:

$$\psi_{k,l}^j(m, n) = \psi\left(\frac{m - 2^j k}{2^j}\right) \psi\left(\frac{n - 2^j l}{2^j}\right) \tag{5.32}$$

Different wavelets have been used to capture the texture features of an image. Commonly used wavelets include *Haar, Mexican hat, Morlet, Daubechies, biorthogonal, symlet, Coiflet,* and *Meyer* wavelet. Both Haar and symlet are special members of Daubechies wavelet family. Haar wavelet has been introduced earlier in Sect. 3.3.1. Figure 5.8 shows the 1D profiles six wavelets from some of the wavelet families, while Fig. 5.9 provides a 3D view for 4 of the wavelets.

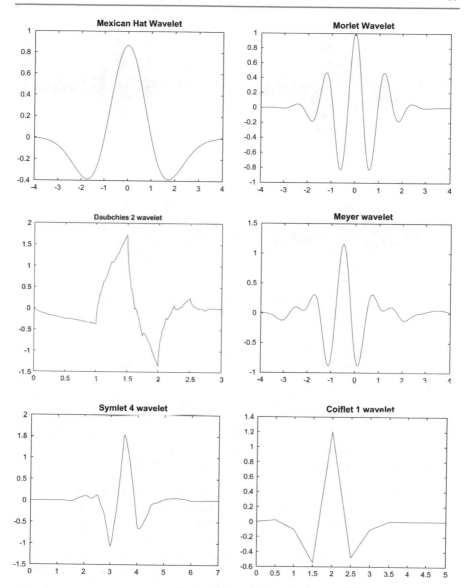

Fig. 5.8 Mexican hat, Morlet, Daubechies, Meyer, Symlet 4, and Coiflet wavelets

Figure 5.10 shows the spectra of the Lena image from some of the common wavelet transforms. It can be seen, that the spectra images are similar but with the subtle difference due to the different shapes of the wavelets. Some are more efficient due to capturing more low-frequency information while discarding more high-frequency information.

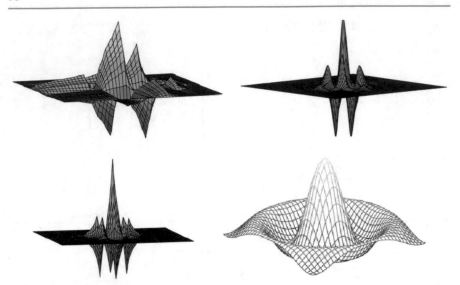

Fig. 5.9 Wavelets in 3D space. Top left: Daubechies 2; Top right: Symlet 4; Bottom left: Coiflet 1; Bottom right: Mexican hat

Discrete wavelets are differentiated by their *shapes*, *orders* (vanishing moments) and *compact support*. The combination of the three factors determines the result of a wavelet transform. The choice of a wavelet usually depends on the nature of the data and applications such as image analysis, image processing, or image compression. Often it requires a number of trials to determine the optimal combination of the three factors. Results of a wavelet transform also depend on whether the wavelet is overlapped or not. The *Maximal Overlap Discrete Wavelet Transform* (MODWT) has found popular application in image analysis.

Shape of a wavelet. The shape of a wavelet is characterized by its *symmetry* and *regularity*. A *symmetric wavelet* shows no preferred direction in time/space, while an asymmetric wavelet gives an unequal weighting to different directions. *Regularity* is related to how many continuous derivatives a function has. Therefore, regularity is a measure of smoothness of a wavelet. Generally, to detect an edge in the data, a wavelet needs to be sufficiently regular. The regularity is also related to the order, the higher the order, the smoother the wavelet.

For image analysis, however, research has shown that the shape of a wavelet does not have a significant influence on classification results [8].

Vanishing moments. An important property of a wavelet function is the number of *vanishing moments*, which characterize how a wavelet interacts with various signals. The names for many wavelets are derived from the number of vanishing moments. For example, db6 is the Daubechies wavelet with six vanishing moments and sym3 is the symlet with three vanishing moments. Generally, a wavelet with N vanishing moments is *orthogonal* to *polynomials* of degree $N - 1$. For example,

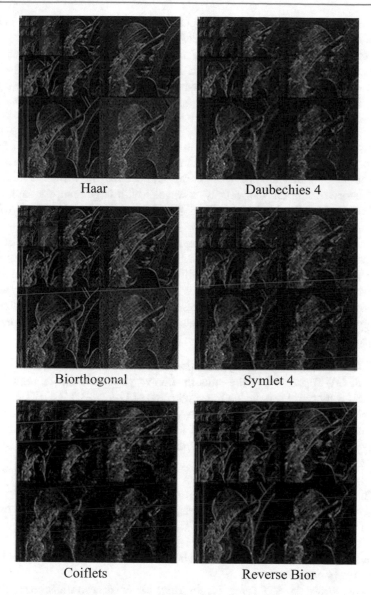

Fig. 5.10 Wavelet spectra for the Lena image

Daubechies 2 wavelet has two vanishing moments. When the Daubechies 2 wavelet is used to transform a data, both the *mean* and any *linear* trend are removed from the data. A higher number of vanishing moments implies that more moments (quadratic, cubic, etc.) will be removed from the data, which results in fewer significant wavelet coefficients. It also means higher order wavelets can capture or represent more frequency bands in the data. Higher order wavelets typically have more oscillations and require wider support. A wavelet with N vanishing moments must have a support of at least length $2N - 1$.

The choice of wavelet order depends on the nature of the data, while usually, it requires a number of empirical tests on the data. Theoretically, the order of the wavelet should be greater than $2H + 1$, where H is the *Hurst exponent* of a signal or data [8, 9]. H can be determined using a similar technique of finding the *fractal dimension* as described in Sect. 5.2.4.

Compact support. This value measures the effective width of the wavelet function. A narrow wavelet function such as haar, db2, or sym2 can capture closely spaced or finer features and are fast to compute, but a narrow wavelet tends to have low-frequency resolution. Conversely, a wavelet with large compact support such as the Daubechies 24 is smoother and has finer frequency resolution which is usually more efficient for denoising.

Wavelets with large support tend to result in coefficients that do not distinguish individual features. Research has shown that wavelets with wider compact support provide increased sensitivity to group differences, which leads to higher classification accuracy [8]. However, wavelet with very large compact support can decrease the localization of prominent features and more coefficients are affected by boundary conditions. In practice, wavelets with an optimal compact support can be found using an empirical test.

MODWT. MODWT is highly redundant and invariant under circular shift. This causes MODWT preserving the smooth time-varying structure in regional time series that is otherwise lost during the application of DWT [8]. MODWT is adaptive to any signal length and emphasizes on variance analysis, which is desirable for feature extraction. Research shows that MODWT has superior performance than ordinary DWT [8]. This is supported by the fact that in literature, Gabor filters are usually preferred than ordinary DWTs.

5.3.3.2   Contrast of DWT and Other Spectral Transforms

If we give a comparison between wavelets, sinusoids (FT), STFT and Gabor filters, we have the following contrasts:

- **Orthogonality**. Both wavelets and sinusoids are orthogonal (1.2–1.6), while Gabor filters are not.
- **Window**. Wavelets, STFT, and Gabor filters are windowed transforms, while FT is not.
- **Window attenuation**. Both wavelets and Gabor filters attenuate towards the border of a window, while STFT does not.
- **Various window size**. Wavelets vary window size, while STFT and Gabor filters do not.
- **Directionality**. Both wavelets and Gabor filters are directional transforms, while FT and STFT are not.

- **Multiresolution**. Wavelets are multiresolution transforms, while Gabor filters, FT, and STFT are not.

Overall, wavelets have more advantage over FT, STFT, and Gabor filters. It can be observed from Fig. 5.10 that different from both FT and STFT, wavelets successfully capture the edge information of the image which is the most useful texture feature. The next is to focus on extracting texture features from the DWT spectrum.

5.3.3.3 Multiresolution Analysis

Space (time)-frequency methods attempt to find a specific frequency at a specific location, which is the main shortcoming of FT and STFT. However, it is not possible to find a specific frequency at a specific location simultaneously, just like we cannot get both a global view (zooming out) and a local view (zooming in) of a map at the same time (we lose global view when getting too close to the map for details and we lose details when we zoom out for a global view of the map). The solution is to use multiresolution or multi-view to create a tradeoff between spatial resolution and frequency resolution.

Multiresolution methods are designed to obtain a good spatial resolution but less accurate frequency resolution at high frequencies or a good frequency resolution but less accurate spatial resolution at low frequencies (Fig. 3.1 left). By using multiresolution or multi-view approach, both global view and local view of an image are obtained and a complete picture of an image is preserved, although not simultaneously. This approach is useful when a signal or an image contains both fine details in small areas and homogenous patches in larger areas. Usually, 2-D images follow this type of frequency patterns. The multiresolution analysis effectively overcomes the window size problem of STFT. Therefore, multiresolution approaches are more effective in image analysis and they overcome the frequency and location dilemma found in both FT (global view) and STFT (local view).

The coefficients at each subband of a wavelet transform are usually scarce, and they are not suitable for direct image representation. Statistics such as those proposed in the GLCM and Gabor filters can be computed from each subband of a wavelet transform. More robust features can be computed from each subband by using Gaussian mixture model. Since high-frequency components are more important for texture representation, features from lower scale subbands are usually given more weight.

Because digital wavelet transform (DWT) is done by two passes of 1D wavelet transform on rows and columns, respectively, wavelets can only capture edge information on horizontal, vertical, and diagonal directions. This gives Gabor filters an edge over wavelet on texture representation and retrieval because Gabor filters can be tuned to more directions than conventional wavelets. However, neither wavelets nor Gabor filters can effectively capture highly anisotropic elements like the curves from an image, and this is the rational behind the introduction of curvelet in the next section.

5.3.4   Texture Features Based on Curvelet Transform

5.3.4.1   Curvelet Transform

Both Gabor filters and wavelets let us do space–frequency analysis of images. Gabor filter is an improvement to STFT by using a Gaussian window and multi-orientations, however, it still uses a fixed window size for different frequencies. This is equivalent to looking at an image for details from a fixed distance, which is difficult. Wavelets use different window sizes for different frequencies, and this creates a multiresolution view of an image and is equivalent to looking for different levels of details of an image from different distance. Therefore, wavelet is a more accurate simulation to human vision system. However, wavelet can only capture texture features from three directions, which are not sufficient for image analysis.

Curvelet [10] has been introduced in literature to take the advantages of both Gabor filters and wavelet, while overcome the limitations of both. Specifically, a curvelet is orthogonal and multiresolution like a wavelet, while it can be tuned to multi-orientation like Gabor filters as well. In other words, a curvelet is a wavelet tuned to multi-orientations and can capture curved or nonlinear edges in an image instead of just linear edges. Therefore, it is a more powerful tool for image analysis.

Basically, curvelet transform extends the ridgelet transform to multiple scale analysis. Given an image $f(x, y)$, the continuous curvelet transform are defined as [11, 12]

$$\Re_f(a, b, \theta) = \iint \psi_{a,b,\theta}(x, y) f(x, y) dx dy. \tag{5.33}$$

where a $(a > 0)$ is the scale, b is the translation, θ is the orientation, and ψ is the curvelet which is defined as follows:

$$\psi_{a,b,\theta}(x, y) = a^{-\frac{1}{2}} \psi \left(\frac{x \cos \theta + y \sin \theta - b}{a} \right) \tag{5.34}$$

where θ is the orientation and a is the scale of the curvelet. A curvelet is constant along the lines: $x \cos\theta + y \sin\theta = b$ and transverse to these ridges are wavelets. Compared with wavelet definition, the location parameters (b_1, b_2) of wavelet are replaced by the line and orientation parameters (b, θ) in a curvelet. In other words, the two transforms are related by [12]

$$\text{Wavelet: } \psi_{scale,\, point\, position}$$

$$\text{Curvelet: } \psi_{scale,\, line\, position}$$

In contrast, Gabor filters are Gaussian-shaped wavelets tuned to different orientations and scales.

(a) **(b)** **(c)**

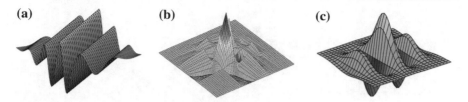

Fig. 5.11 Comparison of curvelet, wavelet and Gabor filter. **a** A curvelet; **b** a Daubechies wavelet and **c** a Gabor filter

$$g_{a,\theta,b_1,b_2}(x,y) = a^{-\frac{1}{2}}g\left(\frac{x\cos\theta + y\sin\theta - b_1}{a}, \frac{-x\sin\theta + y\cos\theta - b_2}{a}\right) \quad (5.35)$$

where the mother wavelet $g(x, y)$ is a Gaussian envelope modulated by a sinusoid wave. The shapes of the three types of wavelets are shown in Fig. 5.11 [11]:

It can be seen from the above figure, compared with both the wavelet and Gabor filter, a curvelet is the most sensitive to lines and edges in an image.

Similar to Gabor filters, a mother curvelet can be tuned to different orientations and different scales to create the curvelets (Fig. 5.12).

Curvelet takes the form of a basis element and obtains a high anisotropy. Therefore, it captures the edge information more effectively because it is sharper than a wavelet and a Gabor filter. Although a curvelet is linear in its edge direction, due to its elongated and orientated design, it aligns with curved edges much better than conventional wavelets do. The contrast between wavelet and curvelet on capturing curved edge information is shown in Fig. 5.13 [12, 13]. It can be observed that curvelets, at higher scales, capture the edge information more accurately and tightly than wavelets.

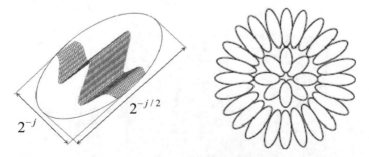

Fig. 5.12 A curvelet and curvelet tiling in spatial domain. Left: a single curvelet with width 2^{-j} and length $2^{-j/2}$; Right: curvelets tuned to 2 scales at different orientations

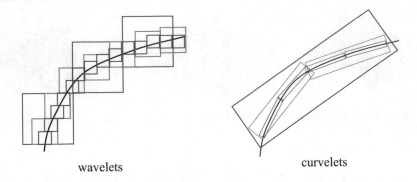

wavelets   curvelets

Fig. 5.13 Edge representation using wavelets and curvelets

5.3.4.2   Discrete Curvelet Transform

Given a digital image $f(m, n)$, $0 \leq m \leq M - 1$, $0 \leq n \leq N - 1$, the discrete
curvelet transform is given as follows:

$$C_{j,\theta}(k, l) = \sum_{n=0}^{N-1} \sum_{m=0}^{M-1} f(m,n)\psi_{j,\theta,k,l}(m,n) \qquad (5.36)$$

where $\psi_{j,\theta,k,l}$ (m, n) is a discrete curvelet; j, θ are the scale and orientation
respectively; and k, l are the spatial location parameters. The frequency response of
a curvelet is a wedge, and the curvelet tiling of frequency plane is shown in
Fig. 5.14.

Curvelets exhibit an oscillating behavior in the direction perpendicular to their
orientation in frequency domain. A few curvelets at different scales and their frequency responses are shown in Fig. 5.15 [3], and the scales shown on the figure are
scales in frequency domain.

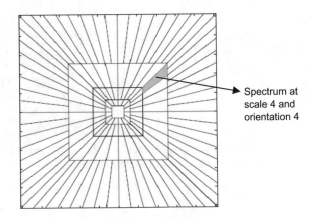

Spectrum at
scale 4 and
orientation 4

Fig. 5.14 Curvelet tiling of frequency plane with 5 level curvelets

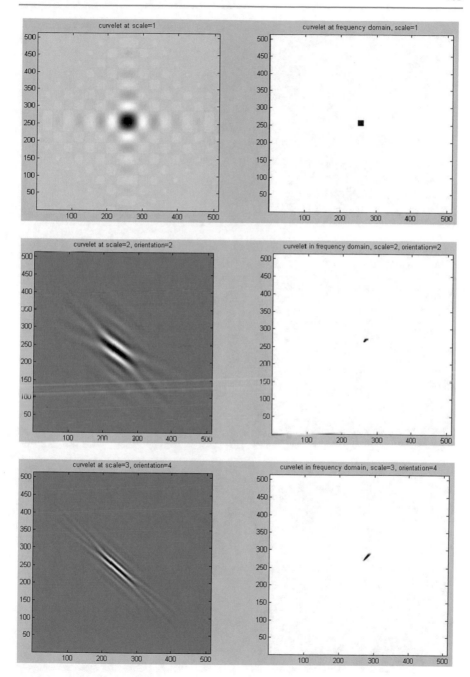

Fig. 5.15 Curvelets at different scales are shown in the spatial domain (left) and in the frequency domain (right) respectively

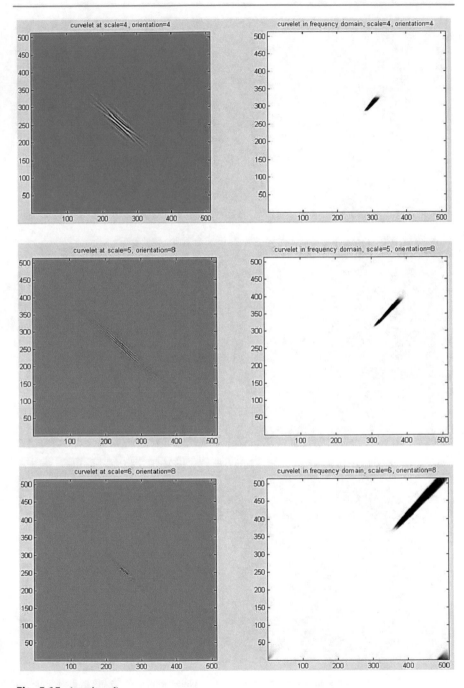

Fig. 5.15 (continued)

It can be observed that the curvelet is nondirectional at the coarsest scale. Whereas, at highest scales, the curvelet waveform becomes so fine that it looks like a needle shaped element. With increase in the resolution level, a curvelet becomes finer and smaller in the spatial domain and shows more sensitivity to curved edges which enables it to effectively capture curves in an image.

Although curvelet has an advantage of capturing nonlinear edges in an image, the computation of curvelet transform is more complex than both the Gabor filters and wavelet. Similar to Gabor filters, to achieve efficiency, curvelet transform is implemented in the frequency domain. That is, both the curvelet and the input image I are transformed into FT domain using FFT and the two FFTs are then multiplied in the FT domain. The product is then transformed back using the inverse fast Fourier transformed (FFT^{-1}) to obtain the curvelet coefficients. The process can be described as

$$\text{Curvelet transform}(I) = \text{FFT}^{-1}[\text{FFT}(\text{curvelet}) \times \text{FFT}(I)] \qquad (5.37)$$

However, due to the FFT of the curvelet is a wedge, the product of the two FFTs needs to be wrapped back into a rectangle before it can be used for the FFT^{-1}. This wrapping process increases the computation cost.

5.3.4.3   Curvelet Spectra

Figure 5.16 shows some of the spectra of the curvelet transform on a flower image at different scales (in frequency domain) and orientations [11, 13]. The size of the spectra is adjusted for better viewing.

To contrast, the spectra of wavelet and Gabor filters are shown in Fig. 5.17.

It can be observed from the above two figures, the spectra of Gabor filters have more redundancy than both the wavelet spectra and curvelet spectra due to the use of overlapping windows during the transform. The spectra of Gabor filters also look more granular than those of both the wavelets and the curvelets. The spectra of wavelets are the most sparse and the most efficient in terms of reducing redundancy, therefore, wavelets are the choice for compression. Curvelet spectra are in between those of wavelet and Gabor filters. Curvelets are more sensitive to edges than Gabor filters and capture edges from more directions than wavelet. Furthermore, curvelets have little redundancy between different subbands. Curvelets also have a complete covering of the spectrum plane and this has overcome the spectra leakage of Gabor filters. However, due to there is a need to wrapping the wedge shape into a rectangle in order to do the invert FFT in (5.37), the computation is more complex than both Gabor filters and wavelets.

5.3.4.4   Curvelet Features

The feature extraction from curvelet transform is similar to Gabor filters. Once the curvelet transform is applied and the coefficients are obtained at each scale and orientation, the mean μ and standard deviation σ are computed for each of the subbands as follows:

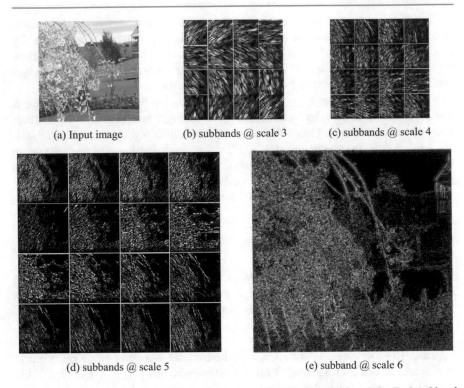

     (a) Input image   (b) subbands @ scale 3   (c) subbands @ scale 4

        (d) subbands @ scale 5   (e) subband @ scale 6

Fig. 5.16 Curvelet subbands at different scales for a flower image (512×512). Each subband captures curvelet coefficients of the input image from one orientation

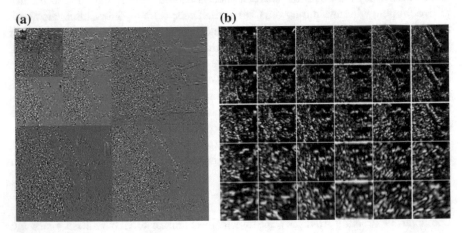

Fig. 5.17 Spectra of wavelets and Gabor filters for the flower image in Fig. 5.16a. **a** Wavelet spectra; **b** Gabor filters spectra

$$\mu_{s\theta} = \frac{E(s,\theta)}{m \times n}, \quad \sigma_{s\theta} = \frac{\sqrt{\sum_x \sum_y \left(|C_{s\theta}(x,y)| - \mu_{s\theta}\right)^2}}{m \times n} \tag{5.38}$$

where s is the scale, θ is the orientation, m and n are the dimensions of the corresponding subband, and $E(s,\theta) = \sum_x \sum_y |C_{s\theta}(x,y)|$ is the total spectral energy of the subband.

Therefore, for each curvelet, two texture features are obtained. If l curvelets are used for the transform, $2l$ texture features are obtained. A $2l$ dimension texture feature vector is used to represent each image in the database for image retrieval. To mitigate the dynamic range of the spectral energy, both the mean and standard deviation features are normalized using the maximum values of the corresponding features in the database.

Based on the curvelet subband division in Fig. 5.14, with 5 levels curvelet decomposition, 82 (= 1 + 16 + 32 + 32 + 1) subbands of curvelet coefficients are computed (only one subband is chosen from the last scale). However, due to the symmetry property, curvelet at angle θ produces the same coefficients as curvelet at angle $\theta + \pi$. Therefore, half of the subbands at scale 2–4 are discarded. As the result, 42 (= 1 + 8 + 16 + 16 + 1) subbands of curvelet coefficients are computed, and a $2 \times 42 = 84$ dimension feature vector is generated for each image. This dimension is higher than the feature vector from Gabor filters with the same scales due to Gabor filters usually use fewer orientations.

Rotation-invariant curvelet features can also be created by using the circular shift method in Gabor filters.

5.3.5   Discussions

Several texture features based on spectral transforms have been introduced in this section, including DCT, Gabor filters, wavelets, and curvelets. Although wavelets are orthogonal and more sensitive to edges, Gabor filters are more directional. This makes Gabor filters a better texture method in many applications. The performance of curvelet based texture features can be affected by the computation complexity due to the irregular frequency response of curvelets.

Generally, spectral texture methods are much more robust than spatial texture methods when they are applied on homogenous texture images, due to their model simplicity and computation efficiency. However, spectral transforms are usually done in a squared window, therefore, it is often difficult to apply spectral texture methods in irregular image regions, especially when the region size is not big enough. In these situations, the DCT-based texture method can be used because the 8×8 window is small enough to fit with most of image regions.

5.4 Summary

This chapter introduces and discusses a number of widely used texture feature extraction methods. Generally speaking, spectral texture methods such as texture features extracted from Gabor filters, wavelet transform, and curvelet transform are much more powerful and robust than spatial methods such as Tamura, GLCM, MRF, and FD. This is because spectral methods are based on either multiresolution or multi-scale analysis which is more robust to noise and scale changes.

However, both spectral and spatial methods have advantages and disadvantages. Spectral texture methods generally capture edge information well while spatial texture methods capture local structures well such as corners, shapes, etc. Spatial texture methods are more intuitive than spectral texture methods. The use of different types of texture methods depends on the applications. For example, for image classification, spectral texture methods are generally preferred; while for object detection and image registration, spatial texture methods are better choices because they need much more accurate matching.

5.5 Exercises

1. Find a grayscale image, use the Matlab code shown in the following web pages to compute the *mean* and *standard deviation* of the image. Then, try them on similar images and different images, explain the pros and cons of the two statistics for image representation.

   https://au.mathworks.com/help/images/ref/mean2.html.

   https://au.mathworks.com/help/images/ref/std2.html.

2. Find a grayscale image and use the Matlab code from the following web page to compute a GLCM of the image: https://au.mathworks.com/help/images/ref/graycomatrix.html. Now calculate the *angular moment*, *contrast*, and *correlation* of the GLCM using Eqs. (5.7–5.9). Try GLCM on the images you used in Exercise 1 and compare the effectiveness of the GLCM statistics with the ordinary statistics from Exercise 1.
3. Find a grayscale image and use the Matlab code from the following web page to compute the spectra of Gabor filters at different scales and orientations. Now, apply the Gabor filters on the same images you used in both Exercise 1 and 2 and compute the mean and standard deviation of the Gabor spectra. Write a short report to compare the effectiveness of Gabor filter statistics with the ordinary statistics and GLCM statistics.

   https://au.mathworks.com/help/images/ref/imgaborfilt.html.

References

1. Tamura H, Mori S, Yamawaki T (1978) Texture features corresponding to visual perception. IEEE Trans Syst Man Cybern 8(6):460–473
2. Islam M (2009) SIRBOT—semantic image retrieval based on object translation. PhD thesis, Monash University
3. Haralick R, Shanmugam K, Dinstein I (1973) Textural features for image classification. IEEE Trans Syst Man Cybern 3(6):610–621
4. Chaudhuri B, Sarkar N (1995) Texture segmentation using fractal dimension. IEEE Trans Pattern Anal Mach Intell 17(1):72–77
5. Rubner Y (1999) Perceptual metrics for image database navigation. PhD thesis, Stanford University
6. Zhang D, Wong A, Indrawan M, Lu G (2000) Content-based image retrieval using Gabor texture features. In: Proceedings of IEEE Pacific-Rim conference on multimedia, Sydney, Australia, 2000
7. Zhang D (2002) Image retrieval based on shape. PhD thesis, Monash University
8. Zhang Z, Telesford Q, Giusti C, Lim K, Bassett D (2016) Choosing wavelet methods, filters, and lengths for functional brain network construction. PLoS ONE 11(6):e0157243
9. Wikipedia, Hurst exponent. https://en.wikipedia.org/wiki/Hurst_exponent. Accessed Feb 2019
10. Starck J, Candès E, Donoho D (2002) The curvelet transform for image denoising. IEEE Trans Image Process 11(6):670–684
11. Zhang D, Islam M, Lu G, Sumana I (2012) Rotation invariant curvelet features for region based image retrieval. Int J Comput Vision 98(2):187–201
12. Do M (2001) Directional multiresolution image representations. PhD thesis, EPFL
13. Sumana I (2008) Image retrieval using discrete curvelet transform. Master thesis, Monash University

Shape Representation

6

The Creator has a model for every creation.

6.1 Introduction

A shape is a binary image. Mathematically, it is defined as

$$f(x, y) = \begin{cases} 1 & if \ (x, y) \in D \\ 0 & otherwise \end{cases} \qquad (6.1)$$

where D is the domain or area of the binary image.

Most of the objects in this world can be identified by their shapes, such as fruits, trees, plant leaves, buildings, furniture, birds, fishes, etc. Figure 6.1 shows a few examples of shape images, the first two are shapes with a contour while the last two are shapes with interior content.

A shape can be defined by its boundary/contour like Fig. 6.1a, b, or by its interior content like Fig. 6.1c, d. There are a variety of shape methods; they can be generally grouped into either contour-based method or region-based method. A number of perceptual shape descriptors have also been proposed to capture both contour and region features. The design of a shape descriptor usually follows the principles suggested by MPEG-7: good retrieval accuracy, compact features, general application, low computation complexity, robust retrieval performance (affine invariance and noise resistance), and hierarchically coarse to fine representation.

However, shape description is a difficult task because it is difficult to define perceptual shape features and measure the similarity between shapes. To make the problem more complex, the shape is often corrupted with noise, defection, arbitrary distortion, and occlusion.

© Springer Nature Switzerland AG 2019
D. Zhang, *Fundamentals of Image Data Mining*, Texts in Computer
Science, https://doi.org/10.1007/978-3-030-17989-2_6

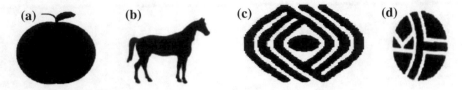

Fig. 6.1 Examples of shape images

6.2 Perceptual Shape Descriptors

There are a number of simple shape descriptors which can be computed according to human perception, such as perimeter, area, compactness, Euler number, circularity, eccentricity, major axis orientation, bending energy, convexity, etc. These individual shape descriptors can be combined to create a more powerful descriptor to represent a shape.

6.2.1 Circularity and Compactness

Circularity represents how a shape is close to a circle, it also indicates how compact the shape is or not. Mathematically, it is defined as the ratio of the area of a shape (A_s) to the area of a circle (A_c) having the same perimeter:

$$C = A_s/A_c \tag{6.2}$$

Assume the perimeter of the shape is p, the area of the circle with the same perimeter p is given by

$$A_c = p^2/4\pi \tag{6.3}$$

Therefore,

$$C = 4\pi A_s/p^2 \tag{6.4}$$

By this definition, the shape with the largest circularity is a circle with circularity of 1. Since 4π is a constant, circularity can be simply defined as

$$C = A_s/p^2 \tag{6.5}$$

Rectangularity and *ellipse variance* can be defined in a similar way to circularity.

The circularity descriptor defined in this way can cause confusions in certain situations. For example, the following two shapes would have the same

Fig. 6.2 Two different shapes with same circularity

circularity or compactness according to the above definition, although perceptually they look very different (Fig. 6.2).

Furthermore, the above circularity is too sensitive to noise and irregularity. Therefore, a more robust circularity descriptor has also been defined. It is defined as the following ratio:

$$C = \sigma_R / \mu_R \tag{6.6}$$

where μ_R stands for the mean of the radial distance from the centroid of the shape to shape boundary points and σ_R stands for the standard deviation of the radial distance from the centroid to shape boundary points.

6.2.2 Eccentricity and Elongation

Eccentricity is defined as the ratio of the length of the longest chord of the shape to the longest chord perpendicular to it. Eccentricity can be computed using either the principle axes method (Fig. 6.3a) or the minimum bounding box method (Fig. 6.3b).

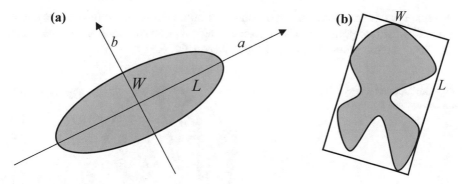

Fig. 6.3 Computation of Eccentricity. **a** Eccentricity with principle axes; **b** eccentricity with minimum bounding box

Fig. 6.4 A curled eel with 0 elongation

In the above figure, the eccentricity of the shapes is given by

$$E = L/W \tag{6.7}$$

The principle axis can also be found using the Principle Component Analysis (PCA) method. *Eccentricity* indicates the *elongation* of a shape, the larger the eccentricity the more elongated the shape.

Elongation is defined as

$$El = 1 - W/L \tag{6.8}$$

$0 \leq El \leq 1$. A circle, a square or any symmetric shape would have the least elongation (0), while objects like eels, poles, road, etc. would have an elongation close to 1. *Elongation*, however, can fail when an elongated shape is bent. For example, a curled eel (Fig. 6.4) would have a small or even 0 elongation although perceptually it is still an elongated object.

6.2.3 Convexity and Solidarity

A region is convex if for any two points with the region, the entire line segment linked by the two points are also inside the region. A convex hull of a shape is the smallest convex region that includes the shape (Fig. 6.5).

Fig. 6.5 A hand shape and its convex hull

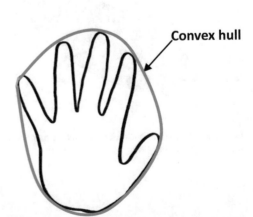

Convex hull

Convexity is then defined as the ratio of the perimeter of the convex hull of the shape (P_h) to the perimeter of the shape (P_s).

$$Convexity = P_h/P_s \tag{6.9}$$

Solidarity is defined as the ratio of the area of the shape (A_s) to the area of its convex hull (A_h).

$$Solidarity = A_s/A_h \tag{6.10}$$

The convexity and solidarity of a convex shape are always 1, convexity and solidarity of non-convex shapes are smaller than 1.

6.2.4   Euler Number

Topology is the study of the properties which are unaffected by any deformation (e.g., rubber-sheet distortion). It is found that when a shape is deformed, the number of holes does not change by the deformation. Therefore, a useful topological descriptor is the *Euler number* which is defined as the difference between the number of holes H and the number of connected components C. A small *Euler number* indicates more holes in a shape.

$$En = C-H \tag{6.11}$$

For example, the Euler numbers of number 3, letter A and B are 1, 0, and -1, respectively (Fig. 6.6).

A Hole Area Ratio (*HAR*) can also be defined in relation to Euler number:

$$HAR = A_h/A_s \tag{6.12}$$

where A_h is the total area of holes and A_s is the area of the shape.

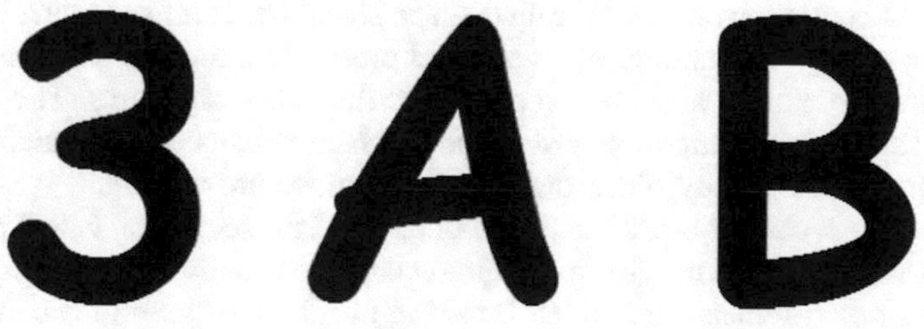

Fig. 6.6 Shapes with different Euler numbers

6.2.5 Bending Energy

The *bending energy* (*BE*) is defined by [1]

$$BE = \frac{1}{N} \sum_{t=0}^{N} K(t)^2$$

and

$$K(t) = (\dot{x}(t)\ddot{y}(t) - \ddot{x}(t)\dot{y}(t))/(\dot{x}^2(t) + \dot{y}^2(t))^{3/2} \qquad (6.14)$$

where $K(t)$ is the curvature function, and N is the number of points on a contour. In order to compute a robust bending energy, the shape boundary is usually Gaussian smoothed before the *BE* calculation. It can be shown that a circle is the shape having the minimum bending energy.

The shape descriptors described in this section are computed according to human perception of the shape patterns. The advantage of using them is that they usually have a semantic meaning. However, the downside of these descriptors is that they are usually sensitive as shown in the circularity and eccentricity sections. It's difficult to describe a shape effectively using a single shape descriptor. Therefore, these perceptual descriptors are usually used as filters to eliminate shapes of large difference. They are often used together with other more powerful descriptors described in the following sections.

6.3 Contour-Based Shape Methods

Contour shape techniques only exploit shape boundary information. There are generally two types of very different approaches for contour shape modeling: *continuous approach* (global) and *discrete approach* (structural) [1]. *Continuous approaches* do not divide shape into subparts, and a multidimensional feature vector derived from the integral boundary is used to describe the shape. It starts to derive a 1D continuous function, called shape signature, from the shape boundary. After that, variety of techniques from signal processing, time series, and statistics can be used to extract a feature vector from the shape signature. The matching between shapes is a straightforward process, which is usually a calculation of the Euclidean distance or city block distance between feature vectors.

Discrete approaches break the shape boundary into segments, called *primitives* using a particular criterion. The final representation is usually a string or a graph (or tree), the similarity measure is done by string matching or graph matching.

6.3.1   Shape Signatures

The first step of contour-based shape methods is to obtain a 1D function from the shape boundary points, called shape signature. Many shape signatures exist, including *complex coordinates, polar coordinates, central distance, tangent angle, cumulative angle, curvature, area,* and *chord length.*

   In general, a *shape signature* $u(t)$ is any 1D function representing 2D areas or boundaries. A shape signature captures the perceptual feature of the shape, it uniquely describes a shape. In the following, we assume the shape boundary coordinates $(x(t), y(t))$, $t = 0, 1, ..., N - 1$, have been extracted in the preprocessing stage, t usually means arclength. The preprocessing usually consists of a denoising procedure or a smoothing procedure and a contour tracing procedure.

6.3.1.1   Position Function

Position function, or *complex coordinates*, is simply the complex number generated from the boundary coordinates:

$$z(t) = [x(t) - x_c] + i[y(t) - y_c] \qquad (6.15)$$

where (x_c, y_c) is the centroid of the shape, which is the average of the boundary coordinates

$$x_c = \frac{1}{N} \sum_{t=0}^{N-1} x(t), \quad y_c = \frac{1}{N} \sum_{t=0}^{N-1} y(t) \qquad (6.16)$$

$z(t)$ is a complex number which captures the spoke features of a shape boundary. $z(t)$ is a translation invariant signature due to the subtraction of the centroid. Rotation causes a circular shift to $z(t)$, and scaling of shape introduces linear change in $z(t)$. The use of position function as shape signature involves little computation. However, the position function needs to be further processed for matching, for example, a centroid distance signature can be computed from $z(t)$.

6.3.1.2   Centroid Distance

The *centroid distance* function is defined as the magnitude of $z(t)$, and it is expressed by the distance of the boundary points to the centroid (x_c, y_c) of the shape [1]

$$r(t) = \left([x(t) - x_c]^2 + [y(t) - y_c]^2 \right)^{1/2} \qquad (6.17)$$

   Same as $z(t)$, $r(t)$ is also invariant to translation. Rotation causes $r(t)$ circular shift and scaling of shape changes $r(t)$ by a linear term. Different from $z(t)$ however, $r(t)$ is a real function which can be used for matching two shapes directly. Figure 6.7 (top) shows the centroid distance signatures of a tree shape. Due to the use of the centroid as the reference point, both $z(t)$ and $r(t)$ attenuate protruding features of a

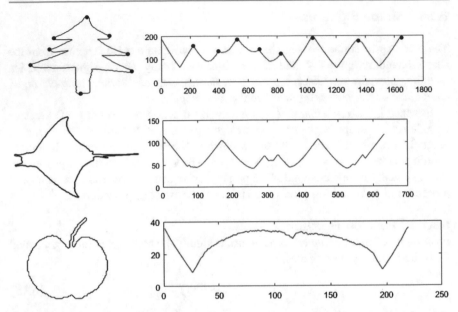

Fig. 6.7 Examples of centroid distance signatures. Top row: a tree shape on the left and its centroid distance signature on the right; middle row: a ray fish on the left and its signature on the right; bottom row: an apple shape on the left and its signature on the right

shape. As can be seen, the $r(t)$ function can generally capture the variations of a shape boundary well, like the eight protruding corners on the tree boundary (marked by black dots). However, it does not capture the most prominent feature in a shape well if it is too thin like the tail of fish or the tail of an apple in Fig. 6.7 (center and bottom).

Another problem with $r(t)$ is that it uses the centroid as reference point. For shapes with high irregularity or low compactness, the centroid often falls outside the shape body. Consequently, the structure of the shape cannot be preserved by $r(t)$ when the shape is under distortion. For example, Fig. 6.8a and b are, respectively, the $r(t)$ functions of two sea snakes, it can be seen that the number of peaks in Fig. 6.8b is doubled compared with that of Fig. 6.8a.

In order to overcome the reference point problem of $r(t)$, *chord length signature* (CLS) $r^*(t)$ has been proposed. The chord length function $r^*(t)$ is derived from shape boundary without using any reference point. For each boundary point $\mathbf{P}(t)$, its $r^*(t)$ is defined as the distance between $\mathbf{P}$ and another boundary point $\mathbf{P}'$ such that $\mathbf{PP}'$ is perpendicular to the tangent vector at $\mathbf{P}$ and $|\mathbf{PP}'|$ is the shortest chord if there are more than one $\mathbf{P}$'s perpendicular to $\mathbf{P}$ (Fig. 6.8c). $r^*(t)$ is invariant to translation. Rotation causes circular shift to $r^*(t)$. Scaling causes linear changes to $r^*(t)$.

$r^*(t)$ is more sensitive to noise than $r(t)$ due to the numeric approximation in computing the tangents at each boundary point and the angles needed to find the chords. To reduce noise sensitivity, an average filter or a median filter can be used to smooth the shape boundary before the signature extraction.

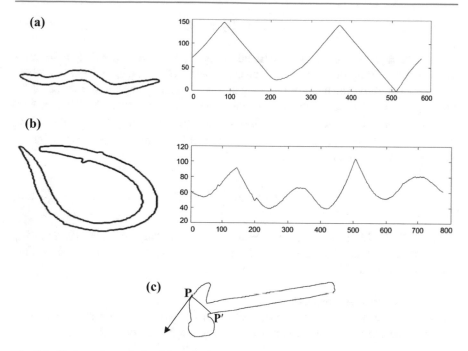

Fig. 6.8 Computation of chord length signature. **a** A sea snake shape on the left and its $r(t)$ function on the right; **b** another sea snake shape on the left and its $r(t)$ function on the right; **c** illustration of computing $r^*(t)$ at point **P** of a hammer shape

6.3.1.3 Angular Functions

Intuitively, the tangent angles of the shape boundary indicate the change of angular directions of the shape boundary. The change of angular directions is important to human perception. Therefore, shape can be represented by its boundary tangent angles

$$\theta(t) = \arctan\frac{y(t) - y(t - w)}{x(t) - x(t - w)} \qquad (6.18)$$

where w, an integer, is a jump step used in practice to smooth the boundary. However, the tangent angle function $\theta(t)$ can only assume values in a range of length π, usually in the interval of $[-\pi/2, \pi/2]$. Therefore, $\theta(t)$ in general contains vertical jump/drop discontinuities at $\theta(t) = \pm\pi/2$ (Fig. 6.9b). One solution is to use the absolute function $|\theta(t)|$ instead of $\theta(t)$, as shown in Fig. 6.9c, the vertical jump/drop discontinuities are removed from $|\theta(t)|$ and the structure of the shape is also preserved. However, the signature function still has sharp corners. Another solution is to use a *cumulative angular function* $\varphi(t)$ which is the net amount of angular bend between the starting position $z(0)$ and position $z(t)$ on the shape boundary

$$\varphi(t) = \theta(t) - \theta(0) \tag{6.19a}$$

The computation of $\varphi(t)$ is illustrated in Fig. 6.9f. As shown in Fig. 6.9d, there is no vertical jump/drop in $\varphi(t)$ at places where $\theta(t) = \pm\pi/2$, and the two sharp angular changes on the heart boundary has been accurately captured. Because of the accumulation, $\varphi(t)$ has captured a linear trend in the function, therefore, a $\psi(t')$ (Fig. 6.9d) can be created by normalizing t into $[0, 2\pi]$ using $t' = \frac{2\pi}{L}t$ and taking away a linear term t' from $\varphi(t')$ if it is obtained in counter clockwise order (or adding t' if it is obtained in clockwise order).

$$\psi(t') = \varphi\left(\frac{L}{2\pi}t'\right) - t' \tag{6.19b}$$

where $t \in [0, L]$, L is the length of the shape boundary and $t' \in [0, 2\pi]$. The cumulative angular signature $\psi(t)$ is invariant to both translation and scaling. Rotation causes a shift in the signature.

6.3.1.4   Curvature Signature

Curvature is an important boundary feature. It is used widely for shape representation in the literature. *Curvature function* is given by (6.20a, 6.20b, 6.20c):

$$\kappa(t) = \frac{d\theta}{dt} \tag{6.20a}$$

$$= \frac{x'y'' - y'x''}{(x'^2 + y'^2)^{\frac{3}{2}}} \tag{6.20b}$$

$$= \frac{\frac{d^2y}{dx^2}}{\left(1 + \left(\frac{dy}{dx}\right)^2\right)^{3/2}} \tag{6.20c}$$

where θ is defined in (6.18). A perfect circular shape would have a constant $\kappa(t)$ based on the definition of (6.20a). Curvature is an important boundary feature, however, due to $\theta(t)$ is typically piecewise continuous and jumps at discontinuities, $\kappa(t)$ is zero almost everywhere and jumps at where $\theta(t)$ jumps. For example, Fig. 6.10 (Top) is the $\kappa(t)$ of the tree shape in Fig. 6.7, while it successfully captures the seven shape corners on the shape boundary, it is jaggy. In order to use $\kappa(t)$ for shape representation, a shape boundary needs to be smoothed before curvature extraction. One way to smooth the shape boundary using a Gaussian filter (Fig. 6.10 (Bottom)).

Curvature is invariant to translation, rotation causes circular shift to the signature, and curvature signature is invariant to scaling if all shapes are normalized to the same number of points.

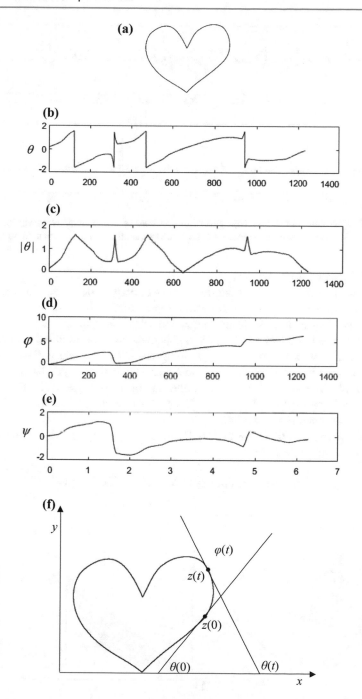

Fig. 6.9 Computation of angular signatures. **a** A heart shape; **b** $\theta(t)$ of (**a**); **c** $|\theta(t)|$ of (**a**); **d** $\varphi(t)$ of (**a**); **e** $\psi(t)$ of (**a**); **f** illustration of the computation of $\varphi(t)$

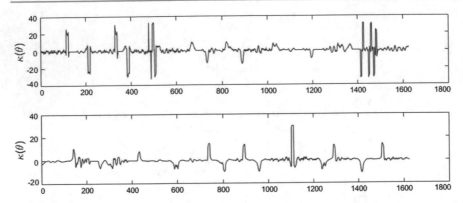

Fig. 6.10 Curvature signatures. Top: curvature signature of a tree shape from Fig. 6.7 without smoothing; Bottom: curvature signature of the same tree shape with a Gaussian smoothing

6.3.1.5   Area Function

It is known that the area of a triangle changes linearly under affine transformation. Linearity is a desirable property for shape representation because normalization of linearity is equivalent to scale normalization which is simple. Therefore, an area signature is used in attempt to acquire a signature invariant to affine distortion [1].

When the boundary points change along the shape boundary, the area of the triangle formed by the two boundary points and the center of gravity also changes (Fig. 6.11a). For each boundary points, the area of the triangle with α degree angle at vertex **o** is calculated (Fig. 6.11b). This forms an *area function* which can be employed as shape representation.

For the triangle ΔOP_1P_2 formed by **O**, P_1, and P_2 in Fig. 6.11b, its area is given by the following difference:

$$\Delta OP_1P_2 = \Delta OP_2x_2 - \Delta OP_1x_1 - \Box x_1P_1P_2 x_2$$

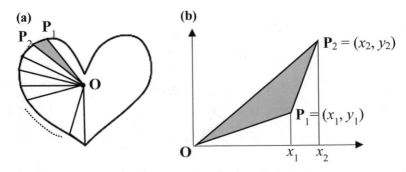

Fig. 6.11 Computation of area signature. **a** Triangulation of a heart shape; **b** illustration on the calculation of the area of the shaded triangle

Mathematically, the area is given by (6.21):

$$A(t) = \frac{1}{2}x_2y_2 - \frac{1}{2}x_1y_1 - \frac{1}{2}(x_2 - x_1)(y_2 - y_1) - (x_2 - x_1)y_1$$
$$= \frac{1}{2}|x_1y_2 - x_2y_1|$$

(6.21)

Because the area of a triangle and the central distance inside the triangle has a linear relationship, $A(t)$ is similar to $r(t)$ [1], however, due to the numerical approximation of the area of a triangle, $A(t)$ is usually more jaggy than $r(t)$. Therefore, $A(t)$ needs to be smoothed for further feature extraction. $A(t)$ is linear under affine transformation.

6.3.1.6 Discussions

Shape signatures reduce shape matching in 2D space into 1D space, this reduces the complexity of feature extraction. Shapes are usually normalized to be translation and scale invariant before signature extraction. Translation invariance is achieved by either using a reference point such as the centroid or using relative positions such as in the cases of curvature, angles and chord length. Scale invariance is achieved by first scaling all shape images to the same size, and then normalizing shape boundaries to the same number of points. If shape signatures are used for shape matching, a shift matching is needed to find the best matching between two shapes. Alternatively, a signature can be quantized into a signature histogram, which is rotation invariant and can be used for matching.

Shape signatures are generally sensitive to noise and irregularities. Therefore, a shape boundary is typically smoothed before signature extraction to remove noise and small irregularities. However, significant irregularities can still cause large error in the matching, therefore, it is impractical to use shape signature for direct representation. Further processing is necessary to improve both matching efficiency and accuracy. The following sections describe methods on feature extraction from shape signatures and other robust contour shape methods.

6.3.2 Shape Context

The idea of shape context is similar to the shape signature methods discussed in Sect. 6.3.1. It also computes contour features from a shape boundary point by point. However, instead of computing a single feature value for each boundary point in the shape signature methods, a feature vector (histogram) is computed in shape context. Furthermore, the computation of the feature vector of each boundary point makes use of all the points on the boundary instead of just two points in the shape signature methods. This makes the shape context features more robust to boundary irregularities. The algorithm of shape context computation is summarized in the following:

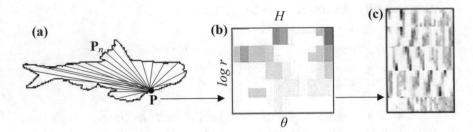

Fig. 6.12 Computation of shape context. **a** A point **P** on a shape boundary and all the vectors started from **P**; **b** the log-polar histogram H of the vectors from **P**, the histogram H is the context of point **P**; **c** the shape context map of shape of (**a**), each row of the context map is the flattened histogram of a point context, the number of rows is the number of sampled points

1. Normalize a shape boundary to N points;
2. For each of the boundary points **P**, find the vectors between **P** and all the other boundary points $\mathbf{P}_n$ (Fig. 6.12a);
3. Quantize both the angles θ and the length r of the vectors at **P**;
4. Create a 2D *logarithmic* histogram H of r on θ for each point **P** (Fig. 6.12b);
5. Flat the 2D histogram H into a 1D histogram h by concatenating the rows of H;
6. Concatenate h's from all the boundary points $\mathbf{P}_n$ to form a histogram map which is the shape context (Fig. 6.12c).

The logarithm of r is to raise the contribution from neighboring points of **P** which would otherwise contribute too little due to significantly shorter vector length than that of points farther away from **P**. The process of computing a shape context is shown in Fig. 6.12 [2]. In Fig. 6.12a, a point **P** on the shape boundary and its vectors to all the boundary points $\mathbf{P}_n$ are shown. Figure 6.12b shows the logarithmic histogram of the vectors in Fig. 6.12a, c shows the shape context map/matrix of the concatenated flatten histograms from all boundary points.

Shape context is invariant to translation due to the use of relative point position. Scaling invariance can be achieved by normalizing all shape boundaries to N points and normalizing the radial distances by the mean distance between all point pairs. Rotation invariance is done by finding the minimum of all shift matching of two shape context maps.

The matching of two shape contexts is complex. It minimizes the total cost of matching between one context matrix and all the permutations of another context matrix. This can affect the robustness of shape context significantly.

6.3.3 Boundary Moments

Moments can be computed to reduce the dimension of a boundary representation. Assume shape boundary has been represented as a *shape signature* $z(i)$, the rth moment m_r and central moment μ_r can be estimated as [1]

$$m_r = \frac{1}{N} \sum_{i=1}^{N} [z(i)]^r \quad \text{and} \quad \mu_r = \frac{1}{N} \sum_{i=1}^{N} [z(i) - m_1]^r \tag{6.22}$$

where N is the number of boundary points. The normalized moments $\bar{m}_r = m_r/(\mu_2)^{r/2}$ and $\bar{\mu}_r = \mu_r/(\mu_2)^{r/2}$ are invariant to shape translation, rotation and scaling.

Boundary moments can also be computed from the boundary histogram. Suppose the amplitude of shape signature function $z(i)$ is quantized and a histogram $p(v_i)$ is created from the quantized $z(i)$. Then, the rth moment is obtained by

$$\mu_r = \sum_{i=1}^{K} (v_i - m)^r p(v_i) \quad \text{and} \quad m = \sum_{i=1}^{K} v_i p(v_i) \tag{6.23}$$

The advantage of boundary moment descriptors is that they are simple to compute and they are more robust than a shape signature. However, only a few low-order moments have physical meaning. In practice, the following three moment descriptors are usually used for shape description: $F_1 = (\mu_2)^{1/2}/m_1$, $F_2 = \mu_3/(\mu_2)^{3/2}$, and $F_3 = \mu_4/(\mu_2)^2$, which describes the variance, skewness, and kurtosis of the boundary.

6.3.4 Stochastic Method

Time-series models and especially autoregressive (AR) modeling have been used for calculating shape descriptors. A linear autoregressive model expresses a value of a function $f(x)$ as a linear combination of a certain number of preceding values. Specifically, each function value in the sequence has some correlation with previous function values and can, therefore, be predicted through a number of, say, M observations of previous function values. The autoregressive model is a simple prediction of the current radius by a linear combination of M previous radii plus a constant term and an error term:

$$f_t = \alpha + \sum_{j=1}^{m} \theta_j f_{t-j} + \sqrt{\beta} \omega_t \tag{6.24}$$

where $\theta_j, j = 1, 2, \ldots, m$ are the AR-model coefficients, m is the model order, it tells how many preceding function values the model uses. $\sqrt{\beta}\omega_t$ is the current error term or residual, reflecting the accuracy of the prediction. α is proportional to the mean of function values. The parameters $\{\alpha, \theta_1, \ldots, \theta_m, \beta\}$ are estimated by using the *least square* (LS) criterion. The estimated $\{\theta_j\}$ are translation, rotation, and scale invariant. Parameters α and β are not scale invariant. But the quotient $\alpha/\sqrt{\beta}$, which reflects the signal-to-noise ratio of the boundary, is regarded as an invariant. Therefore, the feature vector $[\theta_1, \ldots, \theta_m, \alpha/\sqrt{\beta}]$ is used as the shape descriptor.

The AR descriptors can capture the cyclic patterns of shape, it works well for a regular and smooth shape. However, AR is an optimization process, for irregular and complex shapes, it may not have a solution. Furthermore, the choice of m is a complicated problem and is usually decided empirically.

6.3.5 Scale Space Method

6.3.5.1 Scale Space

The problem of noise sensitivity and boundary variations in most spatial domain shape methods inspire the use of scale space analysis. The scale space representation of a shape is created by tracking the position of *inflection points* in a shape boundary or signature. This is done by repeatedly applying low-pass Gaussian filters of variable widths σ at the signature function $f(x)$.

$$L(x; \sigma) = g(x; \sigma) * f(x) \qquad (6.25)$$

The inflection points that remain present in the Gaussian filtered signature functions are expected to be "significant" object characteristics. The result is usually an interval tree, called "fingerprint", consisting of inflection points shown in Fig. 6.12 [3]. As can be seen from Fig. 6.12a, as the scale σ goes up, the number of inflection points on the function $f(x)$ decreases. However, at each point of the function, the inflection point disappears at different scale, this feature has been captured in the interval tree shown in Fig. 6.13b.

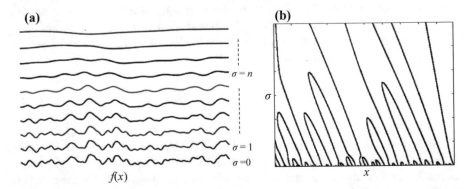

Fig. 6.13 Shape signature in scale space. **a** An original signature function $f(x)$ at the bottom and its successively smoothed versions on the top of it (up to scale 512), where σ is the scale of the smoothed function; **b** the interval tree derived from the zero-crossings of the second derivatives of the smoothed functions at the left, each (x, σ) in the interval tree corresponds to a zero-crossing point at position x and scale σ of the function at left-hand side

6.3.5.2   Curvature Scale Space

The difficulty with scale space method is the interpretation of the interval tree. Mokhtarian and Mackworth [4] developed a *curvature scale space* (CSS) descriptor by finding the peaks of the interval tree. The computation of the CSS descriptor consists of two procedures [5]. The first is to compute a CSS contour map or the interval tree. The second is to extract the branch peaks from the interval tree.

Algorithm of computing CSS contour map:

1. Normalize shape to a fixed number of boundary points;
2. Create an array $ZC[$][ ] to record curvature zero-crossing points at each scale;
3. Set $\sigma = 0$;
4. Compute curvatures of each position at scale σ;
5. Record each curvature zero-crossing point at current scale σ to $ZC[\sigma][x]$;
6. Set $\sigma = \sigma + 1$;
7. Smooth the boundary with a Gaussian filter $g(x; \sigma)$;
8. Repeat step 3–7 until no curvature zero-crossing points are found;
9. Plot $ZC[\sigma][x]$ onto a Cartesian space to create CSS contour map.

Algorithm of extracting CSS contour peaks:

1. Scanning from the top row of CSS contour map;
2. If a zero-crossing point is found at a location (i, j), check the above neighbor points $(i - 1, j - 1)$, $(i - 1, j)$, and $(i - 1, j + 1)$. If the three above neighbor points are nonzero-crossing points, then the location (i, j) is a peak candidate; find all the peak candidates in row i;
3. For each peak candidate (i, j) at row i, check its neighbor peak candidates, if a neighbor candidate (i, k) is found over five points away, then (i, j) is a peak. If a neighbor candidate is found within five points, there is a peak at the middle $(i, (j + k)/2)$;
4. Repeat step 2 and 3 for each row, until all the CSS peaks are found.

 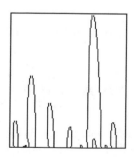 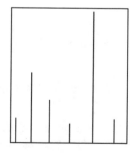

Fig. 6.14 Computation of curvature scale space. A fish shape (left), its CSS contour map (center) and CSS peaks (right)

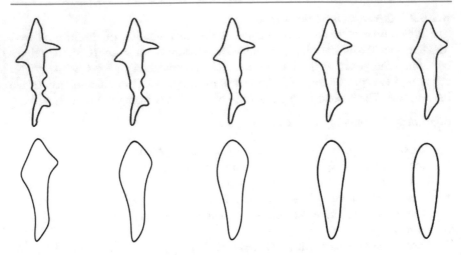

Fig. 6.15 The evolution of shape boundary as scale σ increases

Figure 6.14 shows an example of a fish shape, its CSS contour and peaks [1]. The CSS contour successfully captures the seven major corners of the shape.

The morphing of the fish shape in Fig. 6.14 during the Gaussian smoothing process is shown in Fig. 6.15. As can be seen from Fig. 6.15, as the scale σ goes up, the shape becomes smoother and smoother until it is completely smoothed.

The matching of two CSS descriptors is complex. Assume all shapes have been normalized into the same number of boundary points, the peaks need to be normalized and circularly shifted to find the best match between two shapes. However, due to the sensitivity of both the height and position of the highest peaks, certain tolerance needs to be accepted during the matching of two peaks. The complex matching can affect the performance significantly. In practice, CSS is combined with a few robust single descriptors such as compactness, elongation, etc. [5]. But this introduces a new issue on the weight given to each type of descriptors.

6.3.6 Fourier Descriptor

For any 1D signature function $f(x)$ derived from Sect. 6.3.1, its discrete Fourier transform is given by

$$a_n = \frac{1}{N} \sum_{t=0}^{N-1} f(x) \exp(-j2\pi nx/N), \quad n = 0, 1, \ldots, N-1 \qquad (6.26)$$

This results in a set of Fourier coefficients $\{a_n\}$, which is a representation of the shape. Since shapes generated through rotation, translation, and scaling (called *similarity transformation*) of the same shape are similar shapes, a shape representation should be invariant to these operations. The selection of different start point on the shape boundary to derive $f(x)$ should not affect the representation.

The magnitudes of Fourier coefficients $|a_n|$ are invariant to rotation and starting point because rotation and starting point only affect the phases of the coefficients. Translation invariance can be achieved by normalizing a shape to its center of gravity or subtraction of the mean. Scale invariance is done by normalizing the magnitudes of the coefficients by the DC component a_0, which is the average energy of the signal and is the largest coefficient:

$$|b_n| = |a_n|/a_0, \quad n = 1, 2, \ldots, N-1 \tag{6.27}$$

The set of magnitudes of the normalized Fourier coefficients $\{|b_n|, 1 < n < N - 1\}$ are used as the shape descriptor, denoted as FD: $\{FD_n, 1 < n < N - 1\}$.

FD has several desirable features compared with other shape descriptors. First, it is efficient to compute due to the use of fast Fourier transform or FFT. Second, it provides a coarse to fine representation of a shape. Figure 6.16 examples show how a shape can be reconstructed or represented to a different level of details using different number of Fourier coefficients [1]. In practice, only a small number of low-frequency FDs are used to describe a shape to reduce the sensitivity to noise and irregularities.

Third, all FDs have physical meaning, they capture the different frequency components or different level of details of a shape boundary. Fourth, the matching of two FDs are very simple, it is done by either the city block or Euclidean distance. Because of these desirable features, FD is one of the most robust shape descriptors for contour based shapes. The study by Zhang et al. [5] shows that FD outperforms CSS descriptor which has been adopted by MPEG-7.

(a)

(b)

Fig. 6.16 Reconstructed shapes of an apple shape using Fourier coefficients from **a** $r(t)$; **b** $z(t)$. In both (**a**) and (**b**), from left to right, the shapes are reconstructed using 5, 10, 20, 30, and all the Fourier coefficients, respectively

6.3.7 Discussions

The global shape descriptors described above are basically a mathematical summarization of the boundary samples. However, just like any mathematics, it is based on ideal assumptions and constraints. Typically, these descriptors only work on ideal applications where objects are located in isolation and the boundary of the objects can be viewed or kept in completeness. If the objects overlap each other or certain part of the object boundary is missing, these descriptors do not work anymore. For example, an apple with a bite is still perceived as an apple, like the famous Apple Inc. logo. But using the global shape descriptors, the bite on the apple will change the mathematical summarization dramatically and cause a mismatch. In such kind of applications, *structural shape methods* can be used to analyze the structure of a shape boundary and match shapes using part of the shape boundary. In the following, different types of *structural shape methods* are described in details.

6.3.8 Syntactic Analysis

Syntactic analysis is inspired by the phenomenon that composition of a natural scene is analog to the composition of a language, that is, sentences are built up from phrases, phrases are built up from words and words are built up from alphabets, etc. [6, 7]. In syntactic methods, a shape is represented with a set of predefined primitives. The set of predefined primitives is called the *codebook* and the primitives are called *codewords*. For example, given the codewords in the right of Fig. 6.17, the hammer shape at the left can be represented as a grammatical string of S:

$$S = a\,b\,b\,b\,c\,b\,b\,c\,b\,d\,b\,b \tag{6.28}$$

The matching between shapes can use string matching by finding the minimal number of edit operations to convert one string into another.

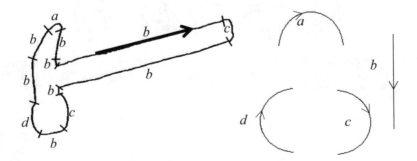

Fig. 6.17 Syntactic analysis of a hammer shape

A more general method is to formulate the representation as a string grammar. Each primitive is interpreted as an alphabet of some grammar, where grammar is a set of rules of syntax that govern the generation of sentences formed from symbols of the alphabet. The set of sentences generated by a grammar, G is called its language and is denoted as $L(G)$. Here, sentences are strings of symbols (which in turn represent patterns), and languages correspond to pattern class. After grammar has been defined, the matching is straightforward. For a sentence representing an unknown shape, the task is to decide in which language the shape represents a valid sentence.

In practice, however, it is difficult to infer a pattern grammar which can generate only the valid patterns. In addition, this method needs a priori knowledge of the database in order to define codewords or alphabets. The knowledge, however, is usually unavailable.

6.3.9   Polygon Decomposition

Polygon can be used to capture the overall shape of a contour and discard the minor variations or noise along the shape boundary. In general, there are two methods to create a polygon from a shape contour: merging and splitting, both are based on applying a distance threshold on the cumulated distance (or errors) between the shape boundary and the polygon line segments [8].

Merging methods add successive pixels to a line segment if each new pixel that is added doesn't cause the segment to deviate too much from a straight line.

In the merging method, it chooses one point as a starting point on the contour, for each new point to be added, let a line segment go from the starting point to this new point. Then, the total squared error of all the boundary points to the line segment is computed. If the error exceeds some threshold, the line from the starting point to the previous point is kept and new line segment is started. For example, in the Fig. 6.18 [8], to find out if the boundary points between $\mathbf{P}_i$ and $\mathbf{P}_k$ should be merged, the total distance or error of all the boundary points to the line segment $\mathbf{P}_i\mathbf{P}_k$ is computed. If the error is larger than the threshold, keep $\mathbf{P}_i\mathbf{P}_j$ and add a new line segment $\mathbf{P}_j\mathbf{P}_k$. The distance d_j from $\mathbf{P}_j$ to $\mathbf{P}_i\mathbf{P}_k$ is given as (6.29) [9]. d_j is equal

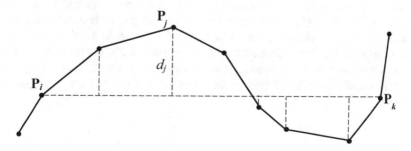

Fig. 6.18 Polygon approximation by merging

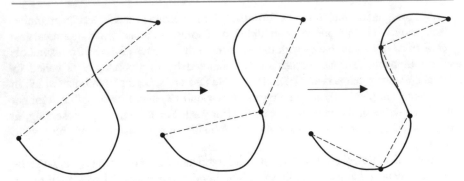

Fig. 6.19 Polygon approximation by splitting

to twice the area of triangle $\Delta \mathbf{P}_i\mathbf{P}_j\mathbf{P}_k$ divided by the distance between $\mathbf{P}_i\mathbf{P}_k$ and the area of $\Delta \mathbf{P}_i\mathbf{P}_j\mathbf{P}_k$ is given in (6.21).

$$d_j = \frac{\left|(x_k - x_i)(y_j - y_i) - (x_j - x_i)(y_k - y_i)\right|}{\sqrt{(x_k - x_i)^2 + (y_k - y_i)^2}} \qquad (6.29)$$

Splitting methods work by first drawing a line from the start point of the boundary to the end point of the boundary. Then, the perpendicular distance from each point along the boundary to the line is computed. If this exceeds some threshold, the boundary is broken into two segments with equal length. The process is repeated for each of the two new segments until no boundary segment needs to be broken (Fig. 6.19).

Once a polygon has been approximated, each segment of the polygon is regarded as a primitive and can be described by its *length* and *angle* in relation to the previous segment. Other features can also be used such as *length ratio* and *triangle area* between two adjacent segments. The polygon is then represented as a string of primitives. The length of each primitive is normalized by the shape boundary to achieve scale invariance.

The matching between two shapes involves shift and best match. Typically, the matching between shapes involves two steps: feature-by-feature matching in the first step and model-by-model matching (shape-by-shape matching) in the second step. In the first step, given a feature(s) of a query shape, the feature(s) is searched through the indexed database, if a particular model feature in the database is found to be similar to the query feature(s), the list of shapes associated with the model feature is retrieved. In the second step, the matching between the query shape and a retrieved model is matched based on the editing distance between the two string of primitives.

Fig. 6.20 Computation of chain code. **a** Chain code in eight-connectivity; **b** chain code in four-connectivity

(a)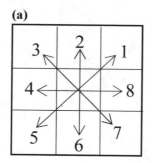

(b)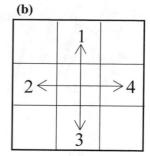

6.3.10 Chain Code Representation

Chain codes describe an object by a sequence of unit-size line segments with a given orientation. In the implementation, a digital boundary of an image is superimposed with a grid, the boundary points are approximated to the nearest grid point, then a sampled image is obtained. From a selected starting point, a chain code can be generated by using a four-connectivity or an eight-connectivity chain code (Fig. 6.20).

Chain code is invariant to translation because the code values are based on directions only. Rotation invariance can be achieved by finding the pixel in the border sequence which results in the minimum integer number, that pixel is then used as the starting pixel. For matching, a *chain code histogram* (of directions) normalized by the chain code length is used. This not only reduces the dimensions and sensitivity to noise but also achieves scaling invariance.

Chain code is itself a fine polygon of equal side length and can be further merged to create a coarse polygon of different side length. In other words, chain code can be used to create a polygon of a shape.

6.3.11 Smooth Curve Decomposition

A contour shape can also be segmented into boundary segments using curvature threshold. The idea is to first smooth the boundary with a Gaussian filter and then calculate the curvature at each point of the smoothed boundary. The boundary is then segmented at points where the curvature exceeds the threshold [10]. The boundary segments are then regarded as the primitives. An example is shown in Fig. 6.21.

The feature for each primitive is its maximum *curvature* and its *orientation*, and the similarity between two primitives is measured by the weighted Euclidean distance. Shapes in database are then indexed with the primitives. Shapes can then be matched using the two steps matching used in the polygon matching.

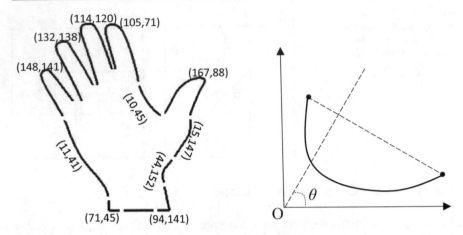

Fig. 6.21 Smooth curve decomposition. Left: a horse shape and its boundary segments; right: the calculation of the angle of a segment

6.3.12   Discussions

The advantage of structural approach is its capability of handling *occlusion* problem in the scene and allowing partial matching. However, structural approach suffers from ambiguity of primitives, expensive computation, and complex matching.

Both global and structural approaches are sensitive to boundary noise and irregularity. This can be overcome by extracting features from shape interior content, or using region-based feature extraction approaches.

6.4   Region-Based Shape Feature Extraction

In region-based methods, all the pixels within a shape region are taken into account to obtain shape features. Common region-based feature extractions methods are based on shape moments. Other region-based methods include grid method, shape matrix, convex hull, and media axis.

6.4.1   Geometric Moments

Mathematically, *geometric moments* are projections of a function onto a polynomial basis in a similar way to the FT which is a projection onto a basis of harmonic sinusoid functions. The geometric moment of order $(p + q)$ of a general function $f(x, y)$ is given as (6.30)

$$M_{pq} = \int\limits_{-\infty}^{\infty} \int\limits_{-\infty}^{\infty} x^p y^q f(x,y) dx dy, \quad p, q = 0, 1, 2, \ldots \qquad (6.30)$$

Two simple properties can be derived from geometric moment:

$$Mass = M_{00} \qquad (6.31)$$

$$\text{Centroid} = \left\{ \bar{x} = \frac{M_{10}}{M_{00}}, \bar{y} = \frac{M_{01}}{M_{00}} \right\} \qquad (6.32)$$

The *central moments* of order $p + q$ of a shape image $f(x, y)$ are given by

$$\mu_{pq} = \sum_x \sum_y (x - \bar{x})^p (y - \bar{y})^q f(x,y) \quad p, q = 0, 1, 2 \ldots \qquad (6.33)$$

The normalized central moments μ_{pq}/μ_{00} are invariant to both translation and scaling. The following properties can be observed from geometric moments.

- $(M_{10}/M_{00}, M_{01}/M_{00})$ defines the *center of gravity* or *centroid* of a shape.
- M_{20} and M_{02} describe the distribution of mass of the shape with respect to the coordinate axes. They are also called the *moments of inertia*.
- μ_{10}/μ_{00} and μ_{01}/μ_{00} are the *horizontal mean* and *vertical mean* of the shape, respectively.
- μ_{20}/μ_{00} and μ_{02}/μ_{00} are the *horizontal variance* and *vertical variance* of the shape, respectively.
- μ_{11} is the *covariance* of the shape.
- μ_{30}/μ_{00} and μ_{03}/μ_{00} represent the *horizontal skewness* and *vertical skewness* of the shape, respectively. The *skewness* measures the *symmetry* of a shape. The skewness of a symmetric shape equals to zero.
- μ_{40}/μ_{00} and μ_{04}/μ_{00} represent the *horizontal kurtosis* and *vertical kurtosis* of the shape, respectively. The kurtosis measures the *peakedness* or *sharpness* of the pixel distribution inside the shape.

These are important properties to describe how pixels are distributed inside the shape. Since both x^p and y^q are monomials, they amplify the moment values of pixels farther away from the centroid, higher order moments reflect how dramatic a shape changes in relation to its centroid.

However, geometric moments are not rotation invariant and it's difficult to derive moment invariants of high order. Hu [11] has derived seven *moment invariants* up to order three:

$$\left.\begin{aligned}
\Phi_1 &= \eta_{20} + \eta_{02} \\
\Phi_2 &= (\eta_{20} - \eta_{02})^2 + 4(\eta_{11})^2 \\
\Phi_3 &= (\eta_{30} - 3\eta_{12})^2 + (3\eta_{21} - \eta_{03})^2 \\
\Phi_4 &= (\eta_{30} + \eta_{12})^2 + (\eta_{21} + \eta_{03})^2 \\
\Phi_5 &= (\eta_{30} - 3\eta_{12})(\eta_{30} + \eta_{12})\left[(\eta_{30} + \eta_{12})^2 - 3(\eta_{21} + \eta_{03})^2\right] \\
&\quad + (3\eta_{21} - \eta_{03})\{\eta_{21} + \eta_{03}\}\left[3(\eta_{30} + \eta_{12})^2 - (\eta_{21} + \eta_{03})^2\right] \\
\Phi_6 &= (\eta_{20} - \eta_{02})\left[(\eta_{30} + \eta_{12})^2 - (\eta_{21} + \eta_{03})^2\right] \\
&\quad + 4\eta_{11}(\eta_{30} + \eta_{12})(\eta_{21} + \eta_{03}) \\
\Phi_7 &= (3\eta_{21} - \eta_{30})(\eta_{30} + \eta_{12})\left[(\eta_{30} + \eta_{12})^2 - 3(\eta_{21} + \eta_{03})^2\right] \\
&\quad + (3\eta_{12} - \eta_{03})(\eta_{21} + \eta_{03})\left[3(\eta_{30} + \eta_{12})^2 - (\eta_{21} + \eta_{03})^2\right]
\end{aligned}\right\} \tag{6.34}$$

where $\eta_{pq} = \mu_{pq}/(\mu_{00})^\gamma$ and $\gamma = 1 + (p + q)/2$ for $p + q = 2, 3, \ldots$.

The geometric moment transform can be extended to generalized form by replacing the conventional transform kernel $x^p y^q$ with a more general kernel of $P_p(x)$ $P_q(y)$.

6.4.2 Complex Moments

The rotation invariance issue of the geometric moments can be easily addressed by using *complex moments* and polar sampling because the magnitude of a complex function is invariant to rotation and polar sampling is also invariant to rotation. The idea is to replace the real polynomials in the geometric moment with complex polynomials and sample the shape in a polar coordinate system. The general form of complex moments is defined as (6.35)

$$C_{pq} = \int_{-\infty}^{\infty} \int_{-\infty}^{\infty} (x + jy)^p (x - jy)^q f(x, y) dx dy \tag{6.35}$$

where $j = \sqrt{-1}$. The orthogonal *Zernike moments* are derived from Zernike polynomials:

$$V_{nm}(x, y) = V_{nm}(\rho \cos \theta, \rho \sin \theta) = R_{nm}(\rho) \exp(jm\theta) \tag{6.36}$$

where

$$R_{nm}(\rho) = \sum_{s=0}^{(n-|m|)/2} (-1)^s \frac{(n - s)!}{s!\left(\frac{n+|m|}{2} - s\right)!\left(\frac{n-|m|}{2} - s\right)!} \rho^{n-2s} \tag{6.37}$$

where ρ is the radius from (x, y) to the shape centroid, θ is the angle between ρ and x-axis, n and m are integers and subject to $n - |m| = even$, $|m| \leq n$. Zernike

Fig. 6.22 The first ten real Zernike polynomials

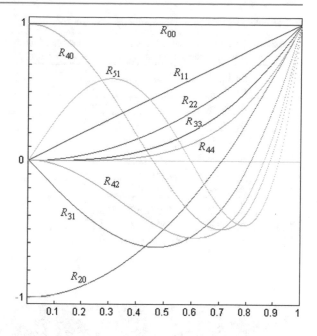

polynomials are a complete set of complex-valued function orthogonal over the unit disk, i.e., $x^2 + y^2 = 1$. Therefore, a shape is first normalized into a unit disk to compute Zernike moments.

For example, the first six real Zernike polynomials are given in the following:

$$R_{00}(\rho) = 1, \quad R_{11}(\rho) = \rho, \quad R_{20}(\rho) = 2\rho^2 - 1$$
$$R_{22}(\rho) = \rho^2, \quad R_{31}(\rho) = 3\rho^3 - 2\rho, \quad R_{33}(\rho) = \rho^3$$

The first 10 real Zernike polynomials of up to order 5 is shown in Fig. 6.22. The complex Zernike moments of order n with repetition m are thus defined as

$$A_{nm} = \frac{n+1}{\pi} \sum_x \sum_y f(x, y) V_{nm}^*(x, y), \quad x^2 + y^2 \leq 1 \tag{6.38}$$

or

$$A_{nm} = \frac{n+1}{\pi} \sum_\rho \sum_\theta f(\rho \cos\theta, \rho \sin\theta) R_{nm}(\rho) \exp(jm\theta), \quad \rho \leq 1 \tag{6.39}$$

where * means complex conjugate. Due to the constraint of $n - |m| = even$ and $m < n$, there are $[n/2]$ repetition of moments in each order n. As an example, the first 36 Zernike moments of up to order 10 are given in Table 6.1. The table only shows the moments with positive m, the moments of negative m are just the rotational versions of the moments of positive m. It can be seen from the table, from

Table 6.1 List of Zernike moments up to order 10

Order (n)	Zernike moment of order n with repetition m (A_{nm})	Number of moments in each order n	Total number of moments up to order 10
0	$A_{0,\,0}$	1	36
1	$A_{1,\,1}$	1	
2	$A_{2,\,0}, A_{2,\,2}$	2	
3	$A_{3,\,1}, A_{3,\,3}$	2	
4	$A_{4,\,0}, A_{4,\,2}, A_{4,\,4}$	3	
5	$A_{5,\,1}, A_{5,\,3}, A_{5,\,5}$	3	
6	$A_{6,\,0}, A_{6,\,2}, A_{6,\,4}, A_{6,\,6}$	4	
7	$A_{7,\,1}, A_{7,\,3}, A_{7,\,5}, A_{7,\,7}$	4	
8	$A_{8,\,0}, A_{8,\,2}, A_{8,\,4}, A_{8,\,6}, A_{8,\,8}$	5	
9	$A_{9,\,1}, A_{9,\,3}, A_{9,\,5}, A_{9,\,7}, A_{9,\,9}$	5	
10	$A_{10,\,0}, A_{10,\,2}, A_{10,\,4}, A_{10,\,6}, A_{10,\,8}, A_{10,\,10}$	6	

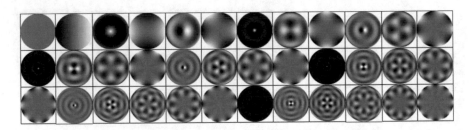

Fig. 6.23 The first 36 Zernike moments from order 1 to 10

the second order (row), there are $[n/2]$ repetition of moments which capture different circular frequencies. The images of Zernike moment functions from order 1 to 10 are shown in Fig. 6.23 [12].

A simplified complex moment is used by MPEG-7, called *angular radial transformation* (ART).

$$ART_{nm} = \frac{1}{2\pi} \sum_{\rho} \sum_{\theta} f(\rho \cos \theta, \rho \sin \theta) V_{nm}(\rho, \theta), \quad \rho \leq 1 \qquad (6.40)$$

where V_{nm} is the ART basis function

$$V_{nm} = R_n(\rho) \exp(jm\,\theta) \qquad (6.41)$$

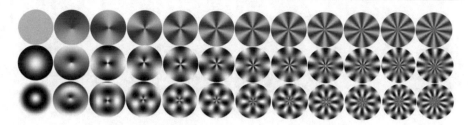

Fig. 6.24 Real parts of the ART basis functions

and $R_n(\rho)$ is the radial basis function

$$R_n(\rho) = \begin{cases} 1 & \text{if } n = 0 \\ 2\cos(n\pi\rho) & \text{if } n \neq 0 \end{cases} \tag{6.42}$$

The real part of the first 36 ART basis functions are shown in Fig. 6.24.

Compared with geometric moments, complex moments are invariant to rotation and they are also more robust due to capturing the spatial information within a shape. They have minimum information redundancy due to orthogonal basis.

However, the computation of complex moments is more expensive than geometric moments. The image needs to be normalized to a unit disk. For an irregular shape, this may either cut out part of the shape if an interior circle is used or include irrelevant part if an exterior circle is used. Depending on how much the shape is cut out or irrelevant part is included, the unit disk normalization can affect the accuracy.

6.4.3   Generic Fourier Descriptor

Complex moment improves the geometric moment with three advantages: rotation invariance, spatial or frequency information, and orthogonality. These three features can be achieved more naturally and efficiently using Fourier transform. The generic Fourier descriptor or GFD is a method just based on this idea [13].

The idea is to first transform a shape into a rectangular polar image with sides r and θ by a polar raster sampling around the shape centroid. Next, a 2D Fourier transform is applied on the transformed rectangular image. The normalized Fourier coefficients are then used as the shape descriptor.

Figure 6.25 demonstrates the polar raster transform [13]. For example, Fig. 6.25a is the original shape image in polar space, Fig. 6.25b is the polar raster sampled image plotted into Cartesian space.

Given a shape image $I = \{f(x, y); 0 \leq x < M, 0 \leq y < N\}$. To apply the polar Fourier transform or PFT, the shape image is converted from Cartesian space to polar space $I_p = \{f(r, \theta); 0 \leq r < R, 0 \leq \theta < 2\pi\}$, R is the maximum radius of the shape. The origin of the polar space is set to be the centroid of the shape so that the shape is translation invariant. The centroid (x_c, y_c) is given by (6.43)

Fig. 6.25 Polar raster transform. **a** An original shape image in polar space; **b** polar raster sampled image of (**a**) plotted into Cartesian space

$$x_c = \frac{1}{M}\sum_{x=0}^{N-1} x, \quad y_c = \frac{1}{N}\sum_{y=0}^{M-1} y \tag{6.43}$$

and (r, θ) is given by

$$r = \sqrt{(x - x_c)^2 + (y - y_c)^2}, \theta = \arctan\frac{y - y_c}{x - x_c} \tag{6.44}$$

The shape image is then polar raster sampled around the centroid and transformed into a rectangular polar image. The polar image, e.g., Figure 6.25b, is a normal rectangular image. Therefore, a 2D FT is applied on this polar rectangle (PFT). The PFT has a similar form to the normal discrete 2D FT in Cartesian space and is defined as

$$PF(\rho, \theta) = \sum_r \sum_i f(r, \theta_i) \exp\left[j2\pi\left(\frac{r}{R}\rho + \frac{2\pi i}{T}\theta\right)\right] \tag{6.45}$$

where $0 \leq r < R$ and $\theta_i = i(2\pi/T)$ $(0 \leq i < T)$; $0 \leq \rho < R$, $0 \leq \theta < T$. R and T are the radial frequency resolution and angular frequency resolution, respectively.

In addition to the three advantages of complex moment, the PFT has another desirable feature of capturing shape information in a smaller number of low-frequency coefficients. This is particularly suitable for shape representation. Figure 6.26 shows an example of PFT on two shape images with different orientations. Conventional 2D FT on the two images results in two different spectra, as they are rotated each other. It can be observed from the middle row of Fig. 6.26 that rotation of shapes in Cartesian space results in circular shift in polar space. However, the circular shift does not change the spectra distribution on polar space, e.g., the bottom row of Fig. 6.26.

Since $f(x, y)$ is a real function, the spectra is circularly symmetric, only the first quarter of the spectra features is needed to describe the shape.

The acquired coefficients of the PFT are translation invariant due to the use of centroid as polar space origin. Rotation invariance is achieved by ignoring the phase information in the coefficients and only retaining the magnitudes of the

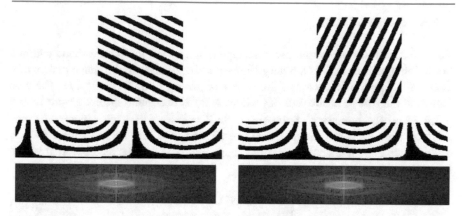

Fig. 6.26 Rotation invariant GFD. Top row: two shape images with different orientations; middle row: the polar raster sampled images of the two corresponding shapes at the top row; bottom row: the Fourier spectra images of the two corresponding images at the middle row

coefficients. To achieve scale invariance, the first magnitude value is normalized by the area of the circle (*area*) in which the polar image resides or the mass of the shape (*mass*), and all the other magnitude values are normalized by the magnitude of the first coefficient. The translation, rotation, and scale normalized PFT coefficients are used as the shape descriptor. To summarize, the shape descriptor derived from the PFT is shown as follows:

$$\mathbf{GFD} = \left\{ \frac{|PF(0,0)|}{area}, \frac{|PF(0,1)|}{|PF(0,0)|}, \cdots, \frac{|PF(0,n)|}{|PF(0,0)|}, \cdots, \frac{|PF(m,0)|}{|PF(0,0)|}, \cdots, \frac{|PF(m,n)|}{|PF(0,0)|} \right\}$$

$$(6.46)$$

where m is the maximum number of the radial frequencies selected and n is the maximum number of angular frequencies selected. m and n can be adjusted to achieve hierarchical coarse to fine representation requirement. Normally, the first coefficient, or the DC component is used as the normalization factor and is discarded after normalization. However, this component is used as an additional feature in a shape descriptor because it reflects the average energy (scale) of the shape which is useful for shape description.

For efficient shape description, only a small number of the acquired GFD features are selected for shape representation. The selected GFD features form a feature vector which is used for indexing the shape.

Compared with Zernike moment descriptor (ZMD), GFD is simpler and more efficient to compute.

6.4.4 Shape Matrix

The most intuitive way to represent a shape is simply to binarize the shape within a bounding box, the result is a binary *shape matrix*. To acquire the shape matrix of a shape S, a square is centered at the *center of gravity* G of S (Fig. 6.27). The side length of the square is equal to $2L$, where L is the maximum distance from G to a point M on the boundary of the shape, or $L = \overline{GM}$. All shape squares are normalized with the same length L and are aligned with line L. The square is then divided into $N \times N$ blocks b_{ij} and the shape matrix is defined as $SM = [c_{ij}]$ where c_{ij} is given by

$$c_{ij} = \begin{cases} 1 & \textit{if } A(S \cap b_{ij}) > A(b_{ij})/2 \\ 0 \end{cases} \tag{6.47}$$

where $A(\cdot)$ is the area function.

It is easy to show that the shape matrix acquired this way is invariant to translation, scale and rotation. The similarity of two shape matrices $A = [a_{ij}]$ and $B = [b_{ij}]$ is given by

$$d(A, B) = 1 - \frac{1}{N^2} \sum_{i=0}^{N} \sum_{j=0}^{N} |a_{ij} - b_{ij}| \tag{6.48}$$

In [14], a binary shape number is created by concatenating the rows of a shape matrix into a vector.

The shape matrix acquired this way is sensitive to boundary noise because the size and orientation of the square bounding box can easily be influenced by the noise. In practice, there needs to be multiple guesses of longest radii, therefore, the best matching of multiple shape matrices is needed.

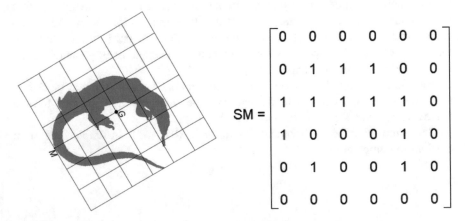

Fig. 6.27 Computation of a shape matrix. A shape on the left and its shape matrix on the right

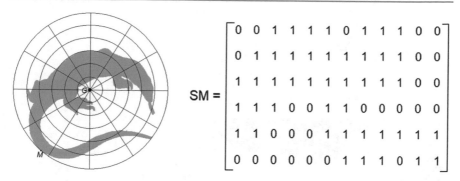

Fig. 6.28 Computation of a polar raster shape matrix. Left: polar raster sampling of a shape; Right: its polar shape matrix

A more robust shape matrix can be created by a using polar grid [15]. The idea is similar to the square shape matrix, however, instead of using a square grid, a polar grid is used. For example, the polar shape matrix of a shape is shown in Fig. 6.28.

The polar model of shape matrix is more robust than the square model because rotation of a shape only causes a horizontal shift of the shape matrix, the matching become simply a shift best matching. However, since the sampling density of the polar raster is not constant at all rings, a weighed shape matrix is necessary.

6.4.5 Shape Profiles

6.4.5.1 Shape Projections

Shape profiles can be extracted from the projections of the shape. The profiles are the projections of the shape onto x-axis and y-axis on the Cartesian coordinate system. By vertical and horizontal projections, two 1D functions are obtained:

$$
\left.\begin{aligned}
P_v(x) &= \sum_{y_{min}}^{y_{max}} f(x, y) \\
P_h(y) &= \sum_{x_{min}}^{x_{max}} f(x, y)
\end{aligned}\right\} \tag{6.49}
$$

The vertical profile $P_v(x)$ counts the number of pixels on each column of the shape and the horizontal profile $P_h(y)$ counts the number of pixels on each row of the shape (Fig. 6.29). The profiles are unique to each type of objects, they can be used as shape signatures to describe shapes.

Polar shape profiles can also be obtained in a similar way by counting the number of pixels at each angle and radius on polar coordinates.

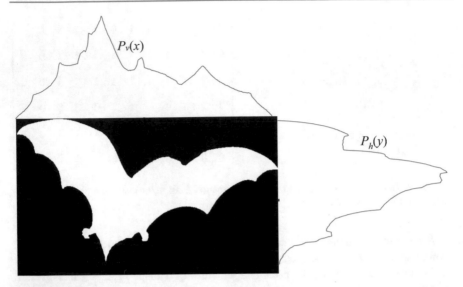

Fig. 6.29 Computation of shape profiles. A binary shape image with its vertical profile $P_v(x)$ (top) and horizontal profile $P_h(y)$ (right-hand side)

6.4.5.2 Radon Transform

Multiple shape profiles projected from different directions θ can be obtained using the *Radon transform*. When all these profiles are aligned on the θ axis in 3D, they create a Radon spectrum of the shape and the spectrum captures the content information of the shape region.

A Radon transform works by creating a shape profile at each angle. Formally, it is defined as

$$R(\rho, \theta) = \iint\limits_{-\infty}^{\infty} f(x, y)\delta(x\cos\theta + y\sin\theta - \rho)dxdy \qquad (6.51)$$

where $\rho = x\cos\theta + y\sin\theta$ is the line the shape is to be projected; θ and ρ are the angle of the line and distance of the line to the origin, respectively. $\delta(x)$ is the Dirac delta-function and is given by (6.51)

$$\delta(x - a) = 0 \quad \text{for} \quad x \neq a \qquad (6.51)$$

$$\int_{-\infty}^{\infty} f(x)\delta(x - a)dx = f(a) \qquad (6.52)$$

The projection of a shape image at a particular angle θ is shown in Fig. 6.30.

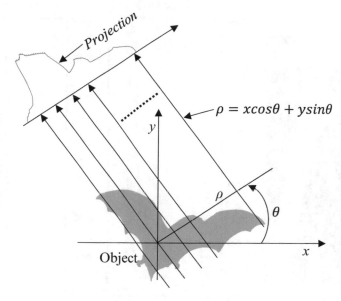

Fig. 6.30 A shape profile from Radon transform

The top row of Fig. 6.31 shows the projections of a hammer shape from two different directions. It can be seen, the projection at 8° has a much higher amplitude than that at 90° due to the projection at 8° captures the profile of the long handle. The two profiles are shown at the bottom row of Fig. 6.31 after zooming in and realignment with the projection angles.

In order to capture the complete information of a shape, projections from all directions are created. If the projections from all directions are created and plotted into an angle-magnitude plane, it creates a spectrum of the transformed shape image. For example, the Radon transform spectrum of the hammer shape is shown in Fig. 6.32. The bright spots on the spectrum indicate high amplitudes on the projections. In this case, there is a single brightest spot at around 8°, it exactly captures the handle angle of the hammer. The next brightest area is around 100°, pointing to the hammerhead. Another Radon transform example is shown in Fig. 6.33, where the most bright spots are at 0° and 90°, pointing to the vertical legs and horizontal body of the dog.

A histogram or a GFD can be computed from the Radon spectrum image as a shape descriptor to match between two shapes.

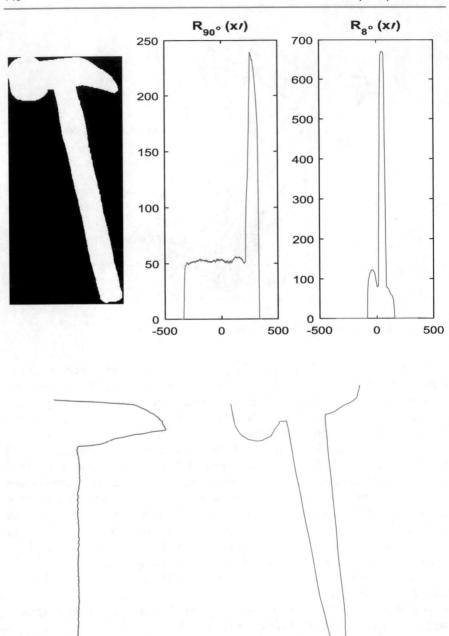

Fig. 6.31 Shape profiles of a hammer image. Top row: A hammer shape and its Radon transforms at 90° and 8°; Bottom row: the 90° and 8° profiles of the hammer shape after zooming in and realignment

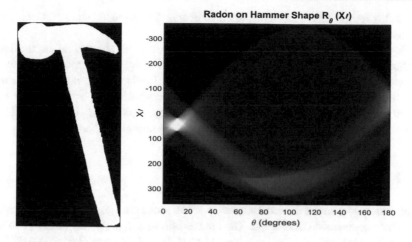

Fig. 6.32 Radon transform of hammer shape. A hammer shape (left) and its Radon transform spectrum (right)

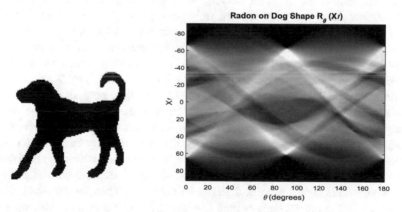

Fig. 6.33 A dog shape and its Radon transform spectrum

6.4.6 Discussions

Global region based methods treat a shape region as a whole instead of just using the boundary, and make effective use of all the pixel information within the region. These methods measure *pixel distribution* within the shape region, which are less likely affected by noise and variations. This makes them more robust than contour-based methods. Methods like complex moments and GFD are more powerful region shape descriptors than conventional moments because they not only capture the pixel distribution within a shape but also capture the spatial details or spatial relationship between pixels. This spatial feature gives them a significant advantage over other region-based methods, such as geometric moments, shape matrix, shape profiles, etc.

However, similar to the global boundary-based methods, global region-based methods cannot deal with overlapped shapes or shapes with missing parts. To address this issue, structural methods are used. Similar to the contour structural methods, *region-based structural methods* decompose a shape region into individual parts which are then used for shape representation and description. In the following, we discuss two of the region-based structural methods.

6.4.7 Convex Hull

A region R is convex if and only if any two points x_1, $x_2 \in R$, the whole line segment $x_1 x_2$ is inside the region. The *convex hull* of a region is the smallest convex region H which satisfies the condition $R \subset H$. The difference $H - R$ is called the *convex deficiency D* of the region R. Methods of computing a convex hull from a shape include morphological methods [16, 17] and polygon approximation. Shape boundaries tend to be irregular because of digitization, noise, and variations in segmentation; these irregularities and noise usually result in a convex deficiency that has small, insignificant components scattered randomly throughout the boundary. Common practice is to first smooth a boundary prior to convex hull computation. The polygon approximation is particularly attractive because it reduces the computation of extracting convex hull from $O(n^2)$ to $O(n)$.

The extracting of convex hull features can be a single process which finds the most significant convex deficiencies along the boundary. The shape can then be represented by a string of concavities. A fuller representation of the shape may be obtained by a recursive process which results in a concavity tree. To do this, the convex hull of an object is first obtained and its convex deficiencies are detected, they are level 1 convex hull and convex deficiencies. Next, the convex hulls and deficiencies of the level 1 convex deficiencies are found. Then, the convex hulls and deficiencies of the level 2 convex deficiencies are found. So on so forth until all the derived convex deficiencies are convex.

Figure 6.34a illustrates the computation of convex hull and concavities [7]. The shape is then represented as a concavity tree in Fig. 6.34b. Each concavity can be described by its area, bridge length which is the line connecting the cut of the concavity, maximum curvature, distance from maximum curvature point to the bridge. The matching between shapes becomes a string matching or a graph matching. A shape boundary needs to be smoothed to remove small irregularities before convex hull and concavity extraction, otherwise, the concavity tree would be very complex and sensitive. For example, if the boundary of the apple shape in Fig. 6.34a had been smoothed, the concavity tree would only have four branches S_1–S_4, which accurately represent the shape boundary.

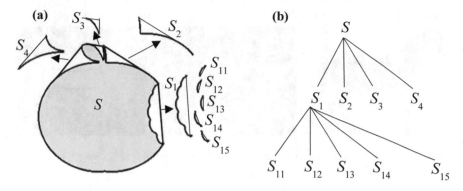

Fig. 6.34 Convex hull and concavity tree of an apple shape. **a** The convex hull of an apple shape and its concavities; **b** Concavity tree representation of the convex hull

6.4.8   Medial Axis

Like the convex hull, region *skeleton* can also be employed for shape representation and description. A skeleton may be defined as a connected set of medial lines along the limbs of a figure [7]. For example, in the case of thick hand-drawn characters, the skeleton may be supposed to be the path traveled by the pen. The basic idea of the skeleton is that eliminating redundant information while retaining only the topological information concerning the structure of the object that can help with recognition.

Shape skeleton can be found by using Blum's *medial axis transform* (MAT) [18]. In MAT, the medial axis is the locus of centers of maximal disks or bi-tangent circles that fit entirely within the shape as illustrated in Fig. 6.35. The bold line in the figure is the skeleton of the hand shape. The skeleton can then be

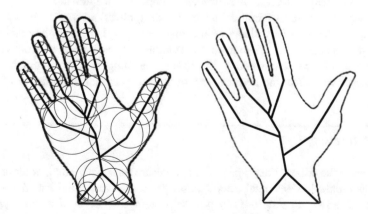

Fig. 6.35 Computation of media axis. Left: construction of medial axis of a rectangle shape using locus of circles; Right: the medial axis or skeleton of the hand shape

(a)   (b)   (c)

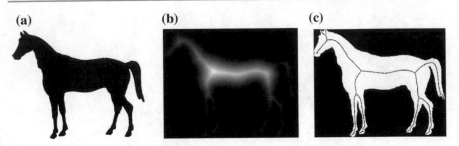

Fig. 6.36 Computation of a shape skeleton. **a** A horse shape image; **b** distance map of (**a**); **c** skeleton of the shape

decomposed into segments and represented as a graph according to certain criteria. The matching between shapes becomes a graph matching problem.

The computation of medial axis in this way is a rather challenging problem. In addition, medial axis tends to be very sensitive to boundary noise and variations. Therefore, it is suggested that the contour of a shape be smoothed before the media axis computation.

The medial axis can be computed from scale space. The medial axis acquired in this way is called the core of the shape [19].

An alternative way of finding the medial axis is to use a distance transform (calculate the distance from each shape point to the background) to convert the binary shape into a gray-level distance map (Fig. 6.36b). A ridge detection is then applied on the distance map followed by a linking process. The ridge points are local extrema which can be found by scanning the distance map both horizontally and vertically. An extrema is found at where the gradient changes from positive (uphill) to negative (downhill). The skeleton of the shape is finally shown up after the linking process (Fig. 6.36c).

The limbs and spine of the skeleton are then detected, features of the limbs and spine such as *length*, *angle*, *curvature* are computed for indexing.

The region structural methods are useful in applications which require partial matching due to object overlapping and missing parts. However, they suffer from similar drawbacks to the contour base structural approaches. Apart from complex computation and implementation, the graph matching is also a complex issue. These issues can affect the performance of region structural methods significantly.

6.5 Summary

This chapter describes three types of shape descriptors: perceptual, contour-based, and region-based. Perceptual shape descriptors are very intuitive and easy to understand, however, they are not powerful enough to be used alone. Typically, they are used as filters to eliminate large number of irrelevant shapes before other shape descriptors are used to refine the retrieval list.

Generally, contour shape descriptors are more sensitive to noise and variations than region shape descriptors due to using less information to extract the features. The sensitivity issue can be alleviated by using spectral transform such as Fourier descriptor. Region-based methods are usually more robust because they make use all the information within the shape region instead of just the boundary.

For image description and retrieval, MPEG-7 has set a set of principles to evaluate the suitability of shape technique: good retrieval accuracy, compact features, general application, low computation complexity, robust retrieval performance, and hierarchical coarse to fine representation. Based on these principles, methods like FD, GFD, ZMD are desirable descriptors.

However, every shape technique has its advantages and disadvantages. The choice of shape techniques often depends on applications and data sets. For example, for shapes with the solid interior, contour-based methods are usually preferred, such as fish sorting, tool recognition, etc. In applications with overlapped objects which require partial matching, structural methods are a preferred choice.

6.6 Exercises

1. Find a binary shape with connected boundary, use the boundary trace code and area code from the following webpages to compute the *circularity/compactness* of the shape. Try more shape images and explain the effectiveness of circularity/compactness.

   https://au.mathworks.com/help/images/boundary-tracing-in-images.html.
   https://au.mathworks.com/help/images/ref/bwarea.html.

2. Find a binary shape image and use the following Matlab code to extract the bounding box of the shape (change *image_name* to your own image name). Then calculate the elongation of the shape.

   ```
   I = imread('image_name');
   h = bwconvhull(I);
   stats = regionprops(h, 'BoundingBox', 'MajorAxisLength', 'MinorAxisLength');
   imshow(I);
   hold on;
   rectangle('Position', stats.BoundingBox, 'EdgeColor', 'b');
   stats
   ```

3. Find a binary image and use the following Matlab code to compute the Radon transform of the image (replace the image *I* with your own image using the *imread()* function shown in the above code): https://au.mathworks.com/help/images/ref/radon.html. Examine the Radon spectrum, match the bright spots on the spectrum and their corresponding features on the binary image.

4. Use the methods shown in Exercise 1 of both Chaps. 4 and 5 to compute the mean, standard deviation and the histogram of the Radon spectrum from Exercise 3.
5. Find more binary shape images, compute the Radon spectra of these images, then compute the statistics and histograms of the Radon spectra. Write a short report on how well the features from Radon spectra describe the images and compare them with the features computed from Exercise 1 and 2.

References

1. Zhang D, Lu G (2004) Review of shape representation and description techniques. Pattern Recognit 37(1):1–19
2. Belongie S, Malik J, Puzicha J (2002) Shape matching and object recognition using shape contexts. IEEE Trans Pattern Anal Mach Intell 24:509–521
3. Stanford University. EE386 Lectures, https://web.stanford.edu/class/ee368/Handouts/Lectures/2018_Winter/13-ScaleSpace.pdf. Accessed Oct 2018
4. Mokhtarian F, Mackworth A (1986) Scale-based description and recognition of planar curves and two-dimensional shapes. IEEE PAMI 8(1):34–43
5. Zhang D, Lu G (2003) Evaluation of MPEG-7 shape descriptors against other shape descriptors. Multimedia Syst 9(1):15–30
6. Fu K (1974) Syntactic methods in pattern recognition. Academic Press
7. Sonka M, Hlavac V, Boyle R (1993) Image processing, analysis and machine vision. Chapman & Hall Computing
8. Yang M, Kpalma K, Ronsin J (2008) A survey of shape feature extraction techniques. Pattern Recognit. 43–90
9. Wikipedia. Distance from a point to a line. https://en.wikipedia.org/wiki/Distance_from_a_point_to_a_line. Accessed Feb 2019
10. Berretti S, Bimbo A, Pala P (2000) Retrieval by shape similarity with perceptual distance and effective indexing. IEEE Trans Multimedia 2(4):225–239
11. Hu M (1962) Visual pattern recognition by moment invariants. IRE Trans Inf Theory IT-8:179–187
12. Fricker P. Pseudo-Zernike Functions, https://au.mathworks.com/matlabcentral/fileexchange/33644-pseudo-zernike-functions. Accessed Feb 2019
13. Zhang D (2002) Image retrieval based on shape. PhD thesis, Monash University
14. Lu G, Sajjanhar A (1999) Region-based shape representation and similarity measure suitable for content-based image retrieval. Multimedia Syst 7(2):165–174
15. Loncaric S (1998) A survey of shape analysis techniques. Pattern Recognit 31(8):983–1001
16. Gonzalez R, Woods R (1992) Digital image processing. Addison-Wesley
17. Davies ER (1997) Machine vision: theory, algorithms, practicalities. Academic Press
18. Blum H (1967) A transformation for extracting new descriptors of shape. In: Whaten-Dunn W (ed) Models for the perception of speech and visual forms. MIT Press, Cambridge, MA, pp 362–380
19. Morse B (1994) Computation of object cores from grey-level images. PhD thesis, University of North Carolina at Chapel Hill

Part III

Image Classification and Annotation

Dripping water penetrates the stone.

Introduction

Due to the rapid digitization and development of the Web, the world is full of digital images. However, without proper classification, these mammoth amount of images are not going to be much helpful; instead, it has caused a huge waste of resources.

A vast amount of research has been done in the past decades to organize digital images into categories so that they can be searched and retrieved conveniently. However, despite the tremendous effort, we are still at the early stage of under-standing images.

Basically, image classification is to organize images into different classes based on the features of the images. Image annotation is to label images with different semantic class names, such as trees, airplanes, lake, etc. The difference between image classification and image annotation is that image annotation attempts to annotate an image with multiple labels or classify an image into multiple classes. Image annotation is done through *Multiple Instance Learning* (MIL). With MIL, an image is represented with a *Bag of Features* (BOF), and an image is labeled as positive if any of the instances in the bag is positive. Image classification and annotation are closely related, because if an image is correctly classified, it can be annotated and if an image is correctly annotated, it can be properly classified into a class.

Given an image as in Fig. III.1, what we want is to classify it into one of the semantic classes, such as "mountain" or "plants" or "nature", with a probability or likelihood.

However, what we have is usually a sequence of numeric features computed through certain feature extraction methods described in Part II, such as a color histogram, or a feature vector as shown in Fig. III.2.

Fig. III.1 An image to be classified into one of the classes

Fig. III.2 An image on the left and it is color histogram on the right

What we can do is to learn from experience or prior knowledge like a human being. Suppose we know the above image is a mountain image, given an unknown image, we can compare its features with the features of the mountain image. If there is a good match (e.g., high probability) between the two feature vectors, we would label or classify the unknown image also as "mountain" (Fig. III.3).

We could use this simple method to identify or retrieve all the mountain images from the database (Fig. III.4). However, this is not going to work well, because the single known image is not a good representation of all the mountain images in the database. Consequently, many mountain images in the database will be misclassified or not retrieved.

A much better way to identify all mountain images in a database is to collect a large number of sample mountain images and use them to train a classifier. Once trained, the classifier will be able to memorize these sample images and use them to recognize unknown images.

There are generally two types of approaches on training or building a classifier, *generative* versus *discriminative*.

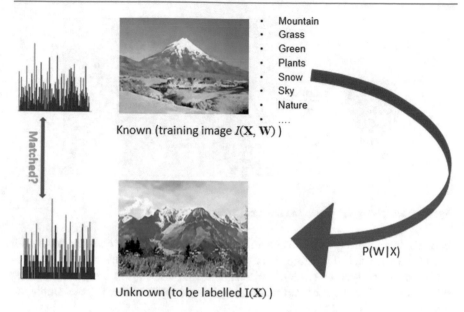

Fig. III.3 Matching between an unknown image with a labeled image

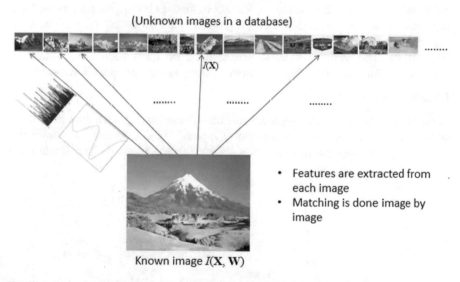

Fig. III.4 Use a labeled image to identify all the mountain images in a database

Generative Model

The generative approach is based on an idea similar to Platonic philosophy that there is an abstract concept or model behind every type of objects in this world, such as trees, apples, dogs, human beings, etc. It is believed that when people try to

learning

learnt
model

training samples

Fig. III.5 Sample apples (right) and the learnt apple model (left)

recognize a specific type of objects in this world, they actually compare them with this abstract model. Therefore, it is possible to create or work out this abstract model for every type or class of objects. The simplest way to create this kind of model is to collect a set of real-world samples and average them, for example, an average apple (Fig. III.5).

However, this can only do simple classification by distinguishing apples from non-fruit objects, while it would be difficult to distinguish apples from other fruits such as peaches, or pears. In practice, a probabilistic distribution is learnt from the collected samples and the distribution is used as the model representing the objects (Fig. III.6). The variety of those probabilistic methods follow the generative approach, including the typical Gaussian mixture model and Bayesian methods.

Discriminative Approach

In contrast to the generative approach, the discriminative approach does not believe or is unaware of the model behind every type of objects. Instead, discriminative methods do classification by comparing different objects based on their similarity or

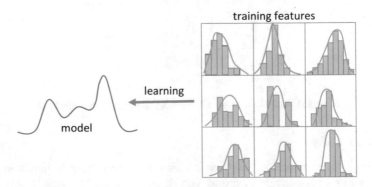

training features

learning

model

Fig. III.6 Computation of a generative model. Sample images are represented as features (right); a mixture distribution model is learnt from those sample features (left)

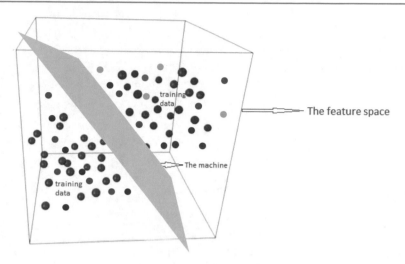

Fig. III.7 A machine is fitted between two classes of data

difference, as opposed to comparing objects with a model in the generative approach. In practice, a large number of sample training data are collected from different classes, and an optimal hyperplane, called the classifier or a machine, is fitted between two classes in a high-dimensional feature space (Fig. III.7). The optimal hyperplane is found through a trial and error process by keeping testing the similarity or difference between the training data.

Bayesian Classification

7

History tells the future.

7.1 Introduction

All Bayesian classification methods are based on the Bayes' theorem which is given below:

$$P(A|B) = \frac{P(B|A)P(A)}{P(B)} = \frac{P(B|A)P(A)}{P(B|A)P(A) + P(B|\bar{A})P(A)} \tag{7.1}$$

where

- A and B are random events and $\bar{A}$ is the complement of A.
- A is a hypothesis to be tested or predicted.
- B is the new data or observation, it is the new evidence to predict A.
- $P(A)$ is called the *prior* probability and $P(B|A)$ is called the *likelihood*, they represent our experience or prior knowledge.
- $P(B)$ is the observation probability or the chance to observe event B.
- $P(A|B)$ is called the *posterior* probability.

The idea of Bayes' theorem is to convert the computation of probability of $P(A|B)$ to $P(B|A)$ which is easier to compute. This is extremely helpful when two random events A and B are dependent on each other and the prediction of event A is difficult due to lack of evidence or information; in this situation, the information about B can be employed to help predicting A more accurately. The information about B can usually be obtained from historical data or experience.

© Springer Nature Switzerland AG 2019
D. Zhang, *Fundamentals of Image Data Mining*, Texts in Computer
Science, https://doi.org/10.1007/978-3-030-17989-2_7

This idea of predicting one event using other related events has been practiced by human beings all the time, for example, we use symptoms to predict a disease, use cloud to predict rain, use a man's culture/background to predict his behavior, use rainfall to predict harvest, etc. In the following, we use two simple examples to get some firsthand understanding of how Bayesian theorem works in real-world applications.

For the first example, we are planning for a sports event at a weekend in a local club and we want to know if the weather will be fine at the weekend. We know the weather and humidity are highly related, we can use humidity to help us predict if the weather is fine so that the sports can go ahead. Formally, let

$$R = \text{Rain} \quad \bar{R} = \text{No rain} \quad H = \text{High humidity (humidity} > 80\%)$$

Suppose we know from history (e.g., Bureau of meteorology) the following prior information:

$$P(R) = 35\% \quad P(\bar{R}) = 65\% \qquad P(H|\bar{R}) = 30\%$$

Suppose further we know there will be high humidity at the weekend based on the most recent day weather, then we can predict the chance of no rain at the weekend by the Bayes' theorem as given below:

$$P(\bar{R}|H) = \frac{P(H|\bar{R})P(\bar{R})}{P(H)} = \frac{P(H|\bar{R})P(\bar{R})}{P(H|\bar{R})P(\bar{R}) + P(H|R)P(R)}$$
$$= \frac{0.3 \times 0.65}{0.3 \times 0.65 + 1 \times 0.65} = \frac{0.195}{0.845} \approx 23\%$$

This information is useful for making a reasonable decision on if the sports event should go ahead. Notice that without the prior information of $P(\bar{R})$ and $P(H|\bar{R})$, the prediction and decision would have been made arbitrarily. In practice, factors such as humidity, temperature, and atmospheric pressure are combined to obtain an even more accurate prediction.

Another example is the application of the Bayes' theorem on image classification. Suppose we have detected some black and white strips (through feature extraction) in an image, based on our experience, we believe it is likely a zebra image. But how likely the image is a zebra, 70% of the chance, 80 or 99%? For many other cases like financial, economic, or military situations, this likelihood is crucial to make a right decision. It is clear we need more evidence to determine the accurate likelihood. The answer is in statistics.

If we were able to sample all the images in the world just like a population census in a country, we would be able to calculate the statistics and tell how many non-zebra images would have the black and white strips. This statistic would help us to determine how likely the image with black and white strips is a zebra image. Unfortunately, we are not able to do a census on all the images in the world, all we can do is to sample part of the image population and create an image database, then use the statistics calculated from the image database to approximate those in the image population.

Formally, let

$$Z = \text{Zebra image} \quad \bar{Z} = \text{non-Zebra image} \quad BWS = \text{Black and white strips}$$

Now, suppose we know the following probabilities from a training set or an image database (our experience or prior information):

$$P(BWS|Z) = 1.0 \quad P(Z) = 0.05 \quad P(BWS|\bar{Z}) = 0.01$$

Then, given the black and white strips in an image, we can predict if there is a zebra in the image by the Bayes' theorem as given below:

$$P(Z \,|BWS) = \frac{P(BWS|Z)P(Z)}{P(BWS)} = \frac{P(BWS|Z)P(Z)}{P(BWS|Z)P(Z) + P(BWS|\bar{Z})P(\bar{Z})}$$
$$= \frac{1.0 \times 0.05}{1.0 \times 0.05 + 0.01 \times 0.95} = \frac{0.05}{0.0595} \approx 84\%$$

Therefore, we can say there is a high chance that the image is a zebra image and we have high confidence to classify the image into the zebra image category.

Bayes' theorem can be extended to multiple events $A_1, A_2, \ldots, A_n$ as follows:

$$P(A_i|B) = \frac{P(B|A_i)P(A_i)}{P(B)} = \frac{P(B|A_i)P(A_i)}{P(B|A_1)P(A_1) + \cdots + P(B|A_n)P(A_n)} \tag{7.2}$$

In this case, B is related to or dependent on multiple other events A_i, and based on what we know about the relationship or dependency between B and each A_i: $P(B|A_i)$, we can predict if a new observation of B is from any of the events A_i.

For example, fever (B) can be caused by many diseases or medical conditions (A_i), such as infection, flu, pneumonia, chickenpox, measles, HIV, meningitis, cancers, malaria, Dengue, etc. However, each disease or medical condition has different chances of causing fever: $P(B|A_i)$. Now, given a patient with fever, (7.2) can be used to determine if the patient has flu: $P(A_i|B)$. In clinical practice, however, symptoms are typically combined to nail a disease, e.g., by combining fever with running nose and headache, flu can be diagnosed with very high accuracy.

7.2   Naïve Bayesian Image Classification

7.2.1   NB Formulation

The *naïve Bayesian* (NB) methods are based on a simple application of the above Bayes' theorem on numerical and high-dimensional image data.

- Given a set of N images: $I = \{I_1, I_2, \ldots, I_N\}$.
- And a set of n semantic classes $C = \{C_1, C_2, \ldots, C_n\}$ (events).
- $I \in C$.
- Each image I is represented by a feature vector $\mathbf{x} = (x_1, x_2, \ldots, x_m)$ (observation).

According to Bayes' theorem, the classification or annotation of image I to class C_i is given by

$$P(C_i|I) = P(C_i|\mathbf{x}) = \frac{P(\mathbf{x}|C_i)P(C_i)}{P(\mathbf{x})} = \frac{P(\mathbf{x}|C_i)P(C_i)}{P(\mathbf{x}|C_1)P(C_1) + \cdots + P(\mathbf{x}|C_n)P(C_n)}$$

(7.3)

or

$$P(C_i|I) = P(C_i|\mathbf{x}) = \frac{P(\mathbf{x}|C_i)P(C_i)}{\sum_{k=1}^{n} P(\mathbf{x}|C_k)P(C_k)}$$

(7.4)

Because the denominator $P(\mathbf{x}) = \sum_{k=1}^{n} P(\mathbf{x}|C_k) \times P(C_k)$ is independent of class C_i ($i = 1, 2, \ldots, n$) and is a constant, (7.4) can be written as follows:

$$P(C_i|I) = P(C_i|\mathbf{x}) = \frac{1}{Z} P(\mathbf{x}|C_i)P(C_i)$$

(7.5)

where $Z = \sum_{k=1}^{n} P(\mathbf{x}|C_k)P(C_k)$ is a scaling factor. The class of image I can be decided using the *maximizing a posterior* (*MAP*) criterion

$$P(C_j|I) = \widehat{C} = \arg\max_i P(C_i|\mathbf{x}) = \arg\max_i \{P(\mathbf{x}|C_i)P(C_i)$$

(7.6)

The prior probabilities $P(C_i)$ is usually uniform for all classes; otherwise, they can be found by the frequency or proportion of samples belonging to class C_i among all classes. Therefore, the classification of image I comes down to modeling the likelihood probability of $P(\mathbf{x}|C_i)$.

Since image features are typically numerical and continuous, they need to be discretized before the likelihood modeling. In practice, the following procedure is used to compute the $P(\mathbf{x}|C_i)$ in (7.6).

Training:

- A training database of images from all the n classes $\{C_1, C_2, ..., C_n\}$ are created.
- Image features from the training database are clustered into m clusters X_j using a certain vector quantization algorithm.
- Next, a cluster centroid $\mathbf{x}_j$ is computed for each of the clusters X_j: $\mathbf{x}_j, j = 1, 2, ..., m$.
- Then, the likelihood $P(\mathbf{x}_j|C_i)$ is calculated by finding the frequency of samples in X_j belonging to class C_i.

$$P(\mathbf{x}_j|C_j) = \frac{No.\ of\ samples\ in\ X_j which\ are\ from\ class\ C_i}{Total\ no.\ of\ samples\ in\ cluster\ X_j} \qquad (7.7)$$

Classification/Annotation:

- Given a new image I with feature $\mathbf{x}$.
- Match feature $\mathbf{x}$ to the closest cluster centroids $\mathbf{x}_j$'s.
- Apply the *MAP* of (7.6) by replacing the likelihood $P(\mathbf{x}|C_i)$ with (7.7) to obtain the posterior probability $P(C_j|I)$.

The classification and annotation of an image with Naïve Bayesian method is illustrated in Fig. 7.1 [1, 2]. There are two major modules in an NB classifier: *training* and *annotation*, and each of the major modules consists of three

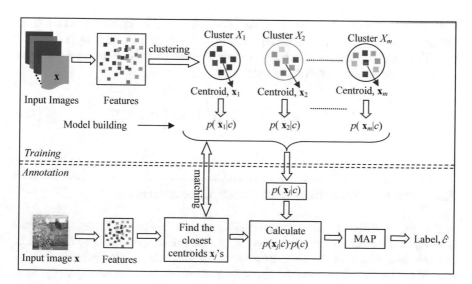

Fig. 7.1 Image classification with Naïve Bayesian method

sub-modules. The training module consists of *feature extraction, clustering,* and *model building,* while the annotation consists of *feature extraction, matching,* and *decision-making* using MAP.

7.2.2 NB with Independent Features

Assume $x_1, x_2, ..., x_m$ are independent of each other, then the likelihood is given as follows:

$$P(\mathbf{x}|C_i) = \prod_{j=1}^{m} P(x_j|C_i) \tag{7.8}$$

There are many situations where the features of a data are independent on each other, e.g., nominal features extracted by a web crawler, such as tree, grass, sand, water, etc. Suppose we are using a set of nominal features $\mathbf{x} = (sand, water, sky, people)$ to classify a collection of images into "beach" and "non-beach" categories, then (7.8) can be employed to modeling the likelihood in the Bayes' theorem. Often different types of numerical image features are combined into a more powerful feature vector, e.g., $\mathbf{x} = (color, shape, texture)$, again, the likelihood probability in the Bayes' theorem can be computed using (7.8).

7.2.3 NB with Bag of Features

If an image I is segmented into k regions, and each region is represented as a feature vector $\mathbf{x}_j, j = 1, 2, ..., k$, I can be represented as a *bag of features*: $I = \{\mathbf{x}_1, \mathbf{x}_2, ..., \mathbf{x}_k\}$. Typically, regions in an image are independent of each other; therefore, the conditional probability of $P(I|C_i)$ is given by

$$P(I|C_i) = P(\mathbf{x}_1, \mathbf{x}_2, ..., \mathbf{x}_k|C_i) = \prod_{j=1}^{k} P(\mathbf{x}_j|C_i) \tag{7.9}$$

7.3 Image Annotation with Word Co-occurrence

In the above naïve Bayesian classification, images are not individually labeled, instead, they are simply classified into categories. The categories can be regarded as *implicit image annotation* or collective image annotation. However, individual images can be pre-labeled and the annotation of images can be done explicitly. Vast amount of labeled images are available on the web, they can be employed to annotate new images. One of the earliest works on *explicit image annotation* or individual image annotation is the *Word Co-occurrence model* (WCC) introduced by Mori et al. [3]. The idea is to establish the relationship between image features and the labels, and use the relationship as a likelihood model to label new images. Specifically, features from pre-labeled image are clustered into clusters and a *word histogram* is computed from each cluster as the likelihood model. The idea of their method can be summarized as follows:

1. Collecting training images with pre-labeled keywords,
2. Divide each image into parts and extract features from each part,
3. Each divided part inherits all words from its original image,
4. Make clusters from all divided images using vector quantization,
5. Accumulate the frequencies of words of all partial images in each cluster, and calculate the likelihood for every word,
6. For an unknown image, divide it into parts, extract their features, and match the image parts with the above clusters. Combine the likelihoods of the image parts and determine which words are most plausible.

The *algorithm* of the word co-occurrence model is given as follows:

- **Collect and label training images**. Given a training dataset of n images $\mathbf{I} = (I_1, I_2, \ldots, I_n)$ and each image I_i is pre-labeled with a set of semantic words $\mathbf{w}_i$:

$$(\mathbf{I}, \mathbf{w}) = \{(I_1, \mathbf{w}_1), (I_2, \mathbf{w}_2), \ldots, (I_n, \mathbf{w}_n)\}$$

- **Obtain the vocabulary of the training images**. The semantic vocabulary of the dataset consists of m words:

$$\mathbf{w} = (w_1, w_2, \ldots, w_m)$$

- **Divide training images into blocks**. Each training image is divided into small blocks and each block inherits all the annotations from its parent image.

- **Vector Quantization (VQ)**

  - Blocks from all the training images are clustered into v clusters represented by the centroids $c_1, c_2, \ldots, c_v$.
  - Each cluster c_i is represented as a feature vector $\mathbf{x}_i$ (each cluster is called a *visual word* or VW, which is corresponding to a region in the training images):

$$\mathbf{c} = (c_1, c_2, \ldots, c_v) = (\mathbf{x}_1, \mathbf{x}_2, \ldots, \mathbf{x}_v)$$

- **Obtain a word histogram in each cluster**. Because each block has inherited a set of words from its parent image, by counting the occurrence of words, a histogram of words from the vocabulary can be created:

$$P(w_j|c_i) = P(w_j|\mathbf{x}_i) = (W_1, W_2, \ldots, W_m) \tag{7.10}$$

where W_j represents the frequency of word w_j in cluster c_i, P(wj |ci) represents the likelihood of word w_j.

- **Annotate an unknown image**

  - Given an unknown image I_u, it is also divided into small blocks and the blocks of the unknown image are also clustered into clusters.
  - Each unknown cluster is matched with the VWs and the nearest l VWs are found for the unknown image.
  - The matching is done by calculating the distance between each feature of the unknown image $\mathbf{x}_u$ and each VW $\mathbf{x}_i$: $\|\mathbf{x}_u - \mathbf{x}_i\|$.
  - The annotation of image I_u to a semantic word w_j ($j = 1, 2, \ldots, k$) is given by first summing up of the histograms of matched clusters c_i ($i = 1, \ldots, l$) and then selecting the top k bins as the annotations:

$$P(w_1, \ldots, w_k|I_u) = \text{top } k \text{ bins} \left(\sum_1^l P(w_j|c_i) \right) \tag{7.11}$$

The co-occurrence annotation method can be illustrated in Fig. 7.2 [1, 2]. There are two key differences between Figs. 7.2 and 7.1. The first is in the training module, while the NB builds a model of $p(\mathbf{x}_i|c)$, the WCC builds a model of $p(w|c)$. The second difference in the annotation module, while the NB makes a decision based on MAP, the WCC makes a decision based on top histogram bins which means an image can be classified into several classes.

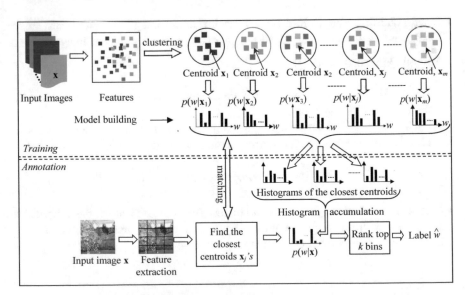

Fig. 7.2 Image annotation with co-occurrence of words

Although the co-occurrence method uses image blocks for the VQ, the blocks can be replaced with pre-segmented image regions. This is because regardless of blocks or regions, they are all represented with a feature vector $\mathbf{x}$, and the VQ is done based on feature vector $\mathbf{x}$.

7.4 Image Annotation with Joint Probability

The word co-occurrence model is a significant development to traditional image classification, it can be generalized into a joint probability model which is described in the following:

- Given a training dataset of n pre-annotated images:

$$(\mathbf{I}, \mathbf{w}) = (I_1, \mathbf{w}_1), (I_2, \mathbf{w}_2), \ldots, (I_n, \mathbf{w}_n)$$

- The semantic vocabulary of the dataset consists of m words:

$$\mathbf{w} = (w_1, w_2, \ldots, w_m)$$

- The annotation or association of an unknown image I to a word w in the vocabulary can be found by the joint probability of $P(w, I)$ or $P(w, \mathbf{x})$, where $\mathbf{x}$ is the feature of I.
- In order to compute $P(w, I)$, a latent variable c is introduced

$$P(w|I) = P(w, I) = P(w|c) \times P(c|I) \tag{7.12}$$

The computation of conditional probabilities $P(w|c)$ and $P(c|I)$ are given in the following procedure:

- The training images are clustered into v clusters or VWs (*the latent variables*):

$$\mathbf{c} = (c_1, c_2, \ldots, c_v) = (\mathbf{x}_1, \mathbf{x}_2, \ldots, \mathbf{x}_v)$$

- An image to be annotated is represented as a histogram or distribution of VWs:

$$P(c_i|I) = (X_1, X_2, \ldots, X_v) \tag{7.13}$$

where X_i is the frequency of VW $\mathbf{x}_i$ in image I

- Each cluster is represented as a histogram or distribution of vocabulary words:

$$P(w_j|c_i) = (W_1, W_2, \ldots, W_m) \tag{7.14}$$

where W_j represents the frequency of word w_j in cluster c_i,

- Finally, the annotation of image I to a word w_j is given by

$$P(w_j|I) = P(w_j|c_i) \times P(c_i|I) \qquad (7.15)$$

As discussed in Sect. 7.2, images can be segmented into regions and represented as a *bag of features* for annotation.

- If an image I is represented as a *bag of features* (pre-clustered): $I = \{\mathbf{x}_1, \mathbf{x}_2, \ldots, \mathbf{x}_k\}$
- The conditional probability of $P(c_i|I)$ in (7.15) can be computed using *MAP*:

$$P(c_i|I) = \arg\max_{c_j} P(c_j|I) = \arg\max_{c_j} \{P(I|c_j) \times P(c_j)\} \qquad (7.16)$$

where

$$P(I|cj) = P(\mathbf{x}_1, \mathbf{x}_2, \ldots, \mathbf{x}_k|c_j) = \prod_{l=1}^{k} P(\mathbf{x}_l|c_j) \qquad (7.17)$$

The key idea of image annotation based on the joint probability model is the association of semantic words with visual words. This is achieved through the VQ process. Once image features are clustered, each cluster (VW) and the semantic words are bound together or associated, because each image feature inherits the semantic word(s) from its parent image. This idea can be illustrated in Fig. 7.3.

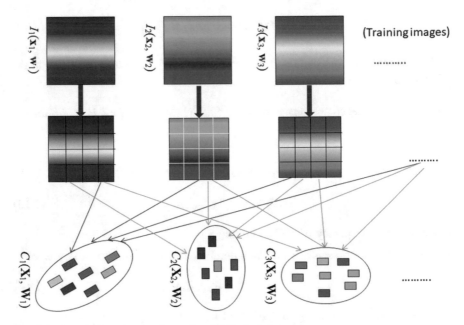

Fig. 7.3 Association of semantic words with block features

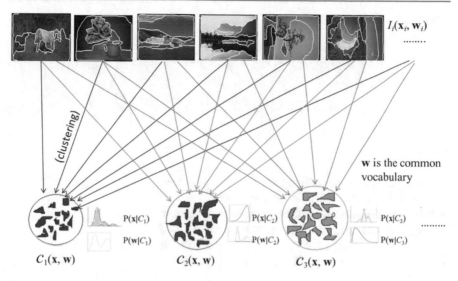

Fig. 7.4 Association of semantic words with region features

Once image features are clustered and visual words are generated, two types of distribution can be created from each cluster: $P(\mathbf{x}|c_i)$ and $P(w|c_i)$. $P(w|c_i)$ is the word distribution in cluster c_i, it connects the VWs to the semantic vocabulary. $P(\mathbf{x}|c_i)$ is the feature distribution in cluster c_i, it connects each VW with each of the images in the database (Fig. 7.4). By combining these two important information, new image can be annotated as shown in the above sections.

7.5   Cross-Media Relevance Model

Although VQ is typical in building the likelihood models, models can also be built "on the fly" by a set of training images which are relevant to the new observation. The *cross-media relevance model* (CMRM) [4] provides an alternative method to the VQ and is another joint probability model.

Given an image which is represented by a set of blobs: $I = \{\mathbf{x}_1, \mathbf{x}_2, \ldots, \mathbf{x}_m\}$, the association of I with concept c is given by the joint probability of $p(c, \mathbf{x}_1, \mathbf{x}_2, \ldots, \mathbf{x}_m)$ as given below:

$$
\begin{aligned}
p(c, \mathbf{x}_1, \cdots \mathbf{x}_m) &= \sum_{J \in T} p(J) \times p(c, \mathbf{x}_1, \cdots \mathbf{x}_m | J) \\
&= p(J) \times \sum_{J \in T} p(c|J) \times \prod_{i=1}^{m} p(\mathbf{x}_i | J)
\end{aligned}
\tag{7.18}
$$

where

$$p(c|J) = (1 - \alpha_J) \times \frac{\#(c, J)}{|J|} + \alpha_J \times \frac{\#(c, T)}{|T|} \qquad (7.19)$$

$$p(\mathbf{x}_i|J) = (1 - \beta_J) \times \frac{\#(\mathbf{x}_i, J)}{|J|} + \beta_J \times \frac{\#(\mathbf{x}_i, T)}{|T|} \qquad (7.20)$$

where

- J is an image in the training set T,
- α_J and β_J are the interpolation parameters,
- $\#(c, J)$ is the number of times concept c appears in J,
- $\#(\mathbf{x}_i, J)$ is the number of times blob $\mathbf{x}_i$ appears in J,
- $\#(c, T)$ is the number of times concept c appears in T, and
- $\#(\mathbf{x}_i, T)$ is the number of times blob $\mathbf{x}_i$ appears in T.

It can be seen from (7.18), given a new observation $I = \{\mathbf{x}_1, \mathbf{x}_2, \ldots, \mathbf{x}_m\}$, CMRM attempts to find all the relevant images in the training set that have both concept $c(p$ $(c|J) \neq 0)$ and feature $\mathbf{x}_i(p(\mathbf{x}_i|J) \neq 0)$. A joint probability model is built by aggregating all the models from the relevant images. This is equivalent to build a class model for concept c "on the fly". From (7.19) and (7.20), it can be seen that the performance of this model depends on the choice of the weights α_J and β_J. In practice, this can be a difficult decision to make.

7.6 Image Annotation with Parametric Model

One of the classic ways of model building is the parametric method using the *expectation-maximization* (EM) algorithm. The idea of image annotation with parametric model is similar to the CMRM method, that is, to build a model for each of the individual images and aggregate the similar individual models into a class model. However, instead of combining training and annotation into a single process as in the CMRM method, parametric model separates training and annotation into two different processes.

During the training, images in the training set are pre-labeled and pre-classified into different classes. A class model is then built by aggregating individual image models in each class. During the annotation, the model of the new image is built and matched with the class models, and the closest classes are selected as the annotations. The procedure of parametric method is shown in Fig. 7.5. The algorithm of this method is as follows:

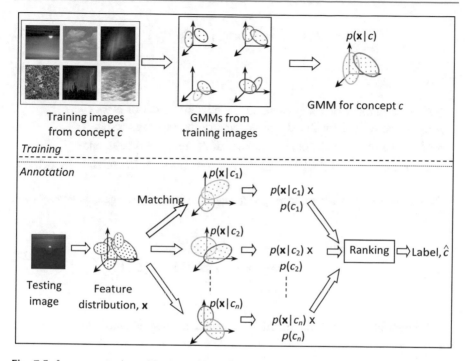

Fig. 7.5 Image annotation with parametric model

- Given a set of N training images: $I_1, I_2, ..., I_N$ and a set of n classes $C_1, C_2, ..., C_n$
- Features (e.g., block features) from each training image I are clustered within the image.
- A Gaussian Mixture Model (GMM) is learned from the clustering using the EM algorithm:

$$P(\mathbf{x}|I) = \sum_{i=1}^{l} \pi_I^i \mathcal{G}\left(\mathbf{x}, \mu_I^i, \Sigma_I^i\right) \tag{7.21}$$

where

- l is the number of components in the mixture model of image I,
- π_I^i is the weight for the ith component of the mixture model,
- μ_I^i is the mean of the ith component of the mixture model, and
- Σ_I^i is the standard deviation of the ith component of the mixture model.
- A Gaussian mixture model for each class C_i is learnt by aggregating (e.g., weighted averaging) all the image models within the class:

$$P(\mathbf{x}|C_i) = \sum_{k=1}^{K} \pi_{C_i}^k \mathcal{G}\left(\mathbf{x}, \mu_{C_i}^k, \Sigma_{C_i}^k\right) \qquad (7.22)$$

where

- K is the number of components in the mixture model of class C_i,
- $\pi_{C_i}^k$ is the weight for the kth component of the mixture model,
- $\mu_{C_i}^k$ is the mean of the kth component of the mixture model, and
- $\Sigma_{C_i}^k$ is the standard deviation of the kth component of the mixture model.

- Given a new observation image $I_u = \mathbf{x}_u$, the annotation of image I_u is given by the *MAP*:

$$
\begin{aligned}
P(C_j|\mathbf{x}_u) = \hat{c} &= \arg\max_{C_i} P(C_i|\mathbf{x}_u) \\
&= \arg\max_{C_i} \{P(\mathbf{x}_u|C_i) \times P(C_i)\}
\end{aligned} \qquad (7.23)
$$

The algorithm of the parametric annotation method is illustrated in Fig. 7.5 [2, 5].

7.7 Image Classification with Gaussian Process

In Gaussian mixture, each multidimensional feature vector $\mathbf{x} = (x_1, x_2, \ldots, x_n)$ is regarded as a data point in a R^n space, and the mixture model is built based on the statistics of the data points in a cluster.

But a multidimensional data $\mathbf{x} = (x_1, x_2, \ldots, x_n)$ can also be regarded as a discretized function $f\colon X \to R$ and $y = f(\mathbf{x}) = \{x_i = f(d_i) \mid i = 1, 2, \ldots, n\}$. A typical example of such a data is a histogram feature vector. Figure 7.6 shows three normalized histograms (vertical bars) from the same class in red, green, and blue, respectively. The corresponding functions approximating the three histograms are shown as colored curves at the top of the histograms.

If we plot all the histogram features $f(\mathbf{x}_i)$ from a class in a single coordinate system, we would see all the data fall within a band and form a cluster. Like in the linear regression which attempts to fit a line to a cluster of data points, we can also fit a curve to this cluster of data points and use this curve as the model to predict new instances. This approach is the idea behind the Gaussian Process or GP, which is demonstrated in Fig. 7.7 [6].

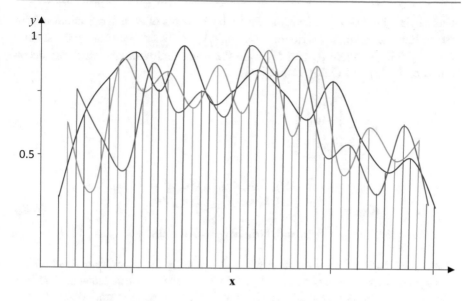

Fig. 7.6 Feature vectors shown as functions. Three histograms shown as vertical color bars and their respective functions shown as colored curves on the top

Fig. 7.7 A cluster of multidimensional data (green) and the approximation function of the data shown in pink

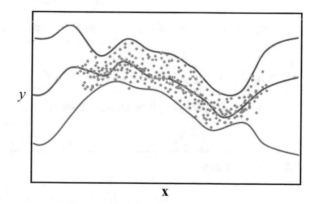

Now, given a set of data $\mathbf{x}_1, \mathbf{x}_2, \ldots, \mathbf{x}_N$ from a certain class C, and each feature vector $\mathbf{x}_i$ is a D-dimensional data point in space $\mathbf{x}_i = (x_{i1}, x_{i2}, \ldots, x_{iD})$. A matrix $\mathbf{X} = D \times N = (\mathbf{d}_1, \mathbf{d}_2, \ldots, \mathbf{d}_D)^\mathrm{T}$ can be created as shown in the following:

$$\mathbf{X} = \begin{bmatrix} x_{11}, & x_{21}, \ldots, & x_{N1} \\ x_{12}, & x_{22}, \ldots, & x_{N2} \\ & \ldots & \\ x_{1D}, & x_{2D}, \ldots, & x_{ND} \end{bmatrix} = \begin{bmatrix} \mathbf{d}_1 \\ \mathbf{d}_2 \\ \ldots \\ \mathbf{d}_D \end{bmatrix} \qquad (7.24)$$

where $\mathbf{d}_j$ is a jth dimensional vector. Since the elements of each $\mathbf{d}_j$ are samples from (or follow) a normal distribution $\mathcal{N}(\mu_i, \sigma_j)$, $\mathbf{X}$ is a Gaussian process and $\mathbf{X} \sim \mathcal{N}(\mathbf{\mu_X}, \mathbf{K_{XX}})$, where $\mathbf{\mu_X}$ and $\mathbf{K_{XX}}$ are the *mean* and *variance* which are determined by (7.25) and (7.26), respectively.

$$
\mathbf{\mu_X} = \begin{bmatrix} \mu(\mathbf{d}_1) \\ \mu(\mathbf{d}_2) \\ \cdots \\ \mu(\mathbf{d}_D) \end{bmatrix} \tag{7.25}
$$

$$
\mathbf{K_{XX}} = \begin{bmatrix} k(\mathbf{d}_1, \mathbf{d}_1), k(\mathbf{d}_1, \mathbf{d}_2) \ldots, k(\mathbf{d}_1, \mathbf{d}_D) \\ k(\mathbf{d}_2, \mathbf{d}_1), k(\mathbf{d}_2, \mathbf{d}_2) \ldots, k(\mathbf{d}_2, \mathbf{d}_D) \\ \cdots \\ k(\mathbf{d}_D, \mathbf{d}_1), k(\mathbf{d}_D, \mathbf{d}_2) \ldots, k(\mathbf{d}_D, \mathbf{d}_D) \end{bmatrix} \tag{7.26}
$$

where $k(\mathbf{d}_i, \mathbf{d}_j)$ is a *kernel* function which is typically the *covariance* function.

To predict a new data or a new set of data $\mathbf{X}_*$, $\mathbf{X}$ and $\mathbf{X}_*$ are concatenated and the concatenated data is a new GP which follows the following normal distribution:

$$
f\begin{pmatrix} \mathbf{X}_* \\ X \end{pmatrix} \sim \mathcal{N}\left(\begin{pmatrix} \mathbf{\mu_{X_*}} \\ \mathbf{\mu_X} \end{pmatrix}, \begin{pmatrix} \sum \mathbf{X_* X_*}, \sum \mathbf{X_* X} \\ \sum \mathbf{XX_*}, \sum \mathbf{XX} \end{pmatrix} \right) \tag{7.27}
$$

Then, the probability of the new data $\mathbf{X}_*$ given the observed data $\mathbf{X}$ is given by (7.28)

$$
p(\mathbf{X}_* | \mathbf{X}) = \mathcal{N}\left(\mathbf{\mu_{X_*}} + \mathbf{K_{X_*X}} \mathbf{K_{XX}^{-1}} (\mathbf{X} - \mathbf{\mu_X}), \mathbf{K_{X_*X_*}} - \mathbf{K_{X_*X}} \mathbf{K_{XX}^{-1}} \mathbf{K_{XX_*}} \right) \tag{7.28}
$$

The proof of (7.28) is given in Appendix [7–9].

7.8 Summary

This chapter introduces the first image classification method: Bayesian classification. Several important and interesting applications of Bayesian classifier are described and demonstrated in details, including NB, word co-occurrence model, CMRM, parametric model, and Gaussian process method. The key features of Bayesian classifiers can be summarized as follows:

1. **Generative**. A Bayesian classifier is a typical generative model, it assumes the distribution or model of likelihood probability is known. This likelihood probability model is typically obtained through learning from known samples.
2. **Intuitive**. Compared with many other black-box-based classifiers such as SVM and ANN, Bayesian classifiers are intuitive, results are easily interpreted by a human being. The basic idea of the Bayesian method is to use our prior

experience to forecast or predict new events. In this sense, we all make decision using Bayesian classifiers.

3. **Robust**. A Bayesian classifier generates a result in probabilistic form instead of deterministic form. Probabilistic prediction is more robust than deterministic prediction, e.g., a 70% chance of rain forecast is more likely to be correct than a rain/no-rain forecast.

4. **Nonlinear**. The boundary of a Bayesian classifier is nonlinear because the prediction is based on the data distributions and the distribution models can be of any shape.

However, the downside of Bayesian classifier is that there needs a large number of data samples to have a reasonable accurate estimation of data distribution.

7.9   Exercises

1. Use the example code in the following web page to generate a Gaussian mixture model: https://au.mathworks.com/help/stats/gmdistribution.html#mw_4758a58e-5bc7-4eda-b261-83521d63d1ce. First, try the gmdistribution function with the following code. Turn the graph to different angles and also to flat (2D) to view the model in more details. Then try different mu and sigma values to create more GMM models.

```
mu = [1 2;-3 -5];
sigma = cat(3,[2 .5],[1 1]);
gm = gmdistribution(mu,sigma);
ezsurf(@(x,y)pdf(gm,[x y]),[-10 10],[-10 10]);
```

2. Use the example code from this link: https://au.mathworks.com/help/stats/fitgmdist.html and try the fitgmdist function with the following code. Turn the graph to different angles and also to flat (2D) to view the model in more details. Now, try fitgmdist with more components and different mu and sigma values to create more complex GMM models.

```
mu1 = [1 2];
sigma1 = [2 0; 0 .5];
mu2 = [-3 -5];
sigma2 = [1 0; 0 1];
rng('default')
r1 = mvnrnd(mu1,sigma1,1000);
r2 = mvnrnd(mu2,sigma2,1000);
X = [r1; r2];
gm = fitgmdist(X,2)
ezsurf(@(x,y)pdf(gm,[x y]),[-8 6],[-8 6])
```

3. Use the code from the following web page to compute the posterior probabilities of a GMM model which you have generated from the above exercises. Explain the graph using the colors and color bar. Write a report on the GMM models, the posterior probabilities and tell how they can be used for image analysis (hints: an image or an image region is a GMM model, and a GMM model is characterized or defined by its parameters e.g., mu and Sigma). https://au.mathworks.com/help/stats/gmdistribution.posterior.html.

References

1. Islam M (2009) SIRBOT—semantic image retrieval based on object translation. PhD thesis, Monash University
2. Zhang D, Islam M, Lu G (2012) A review on automatic image annotation techniques. Pattern Recogn 45(1):346–362
3. Mori Y, Takahashi H, Oka R, Image-to-word transformation based on dividing and vector quantizing images with words. In: Proceedings of the 1st international workshop on multimedia intelligent storage and retrieval management. ACM Press
4. Jeon J, Lavrenko V, Manmatha R (2003) Automatic image annotation and retrieval using cross-media relevance models. In: Proceedings of the 26th international ACM SIGIR conference on research and development in information retrieval, pp 119–1262003
5. Carneiro G, Chan A, Moreno P, Vasconcelos N (2007) Supervised learning of semantic classes for image annotation and retrieval. IEEE PAMI 29(3):394–410
6. Snelson E, Tutorial: Gaussian process models for machine learning. http://mlg.eng.cam.ac.uk/zoubin/tut06/snelson-gp.pdf. Accessed Feb 2019
7. Wikipedia, Covariance matrix. https://en.wikipedia.org/wiki/Covariance_matrix. Accessed Feb 2019
8. Macro, Deriving the conditional distributions of a multivariate normal distribution. https://stats.stackexchange.com/questions/30588/deriving-the-conditional-distributions-of-a-multivariate-normal-distribution. Accessed Feb 2019
9. Wang R, Marginal and conditional distributions of multivariate normal distribution. http://fourier.eng.hmc.edu/e161/lectures/gaussianprocess/node7.html. Accessed Feb 2019

Support Vector Machine

8

To see better, go higher.

One of the key development in recent artificial intelligence (AI) research is the SVM, which has attracted a large amount of research and has produced good results in many applications. Because of the so-called "kernel trick", it has made SVM one of the most effective and efficient machine learning tools in the literature. In this chapter, we attempt to do an anatomy of SVM so that readers have a good understanding of its mechanism.

SVM is basically the combination of both a linear classifier and a k nearest neighbors classifier (K-NN). Therefore, to understand SVM, we will first introduce the linear classifier and the K-NN classifier.

We will only focus on two-class classification problem in this chapter. Because, any classification problem can be converted into a one-vs-all classification, which is a two class classification problem.

8.1 Linear Classifier

The Bayesian methods in Chap. 7 are model based, and they can give good decision if the models are accurate. However, since data distributions are usually unknown, the models can only be estimated accurately if a large number of training samples are available. This is especially true when the number of features is large, which is common for multimedia data.

An alternative approach is to assume that there exists a functional form decision boundary between each pair of classes, and the parameters of the decision boundary or discriminant function can be estimated using available training samples. A linear classifier is one of those approaches.

© Springer Nature Switzerland AG 2019
D. Zhang, *Fundamentals of Image Data Mining*, Texts in Computer Science, https://doi.org/10.1007/978-3-030-17989-2_8

Suppose the data is represented as a n dimensional feature vector $\mathbf{x} = (x_1, x_2, \ldots, x_n)$, then a linear discriminant function is formulated as

$$f(\mathbf{x}) = w_0 + w_1 x_1 + w_2 x_2 + \cdots + w_n x_n \qquad (8.1)$$

where x_i are the variables and w_i are the coefficients or weights. Assume $x_0 = 1$, $f(\mathbf{x})$ can be written as

$$f(\mathbf{x}) = \sum_{j=0}^{n} w_j x_j \qquad (8.2)$$

Geometrically, $f(\mathbf{x}) = 0$ is a hyperplane in n-dimensional space and $f(\mathbf{x}) = 0$ is the decision boundary between two classes. A sample data with feature vector $\mathbf{x}$ is classified into one of the classes using the following criterion:

$$\mathbf{x} \in \begin{cases} class\, 1 & if\, f(x) > 0 \\ class\, 2 & if\, f(x) < 0 \end{cases}$$

8.1.1 A Theoretical Solution

The next is to find out the parameters or the set of weights of $f(\mathbf{x})$: $w_0, w_1, w_2, \ldots, w_n$, which will minimize the number of misclassified samples in a given training set.

A classical way to find the weights is to solve a set of linear equations given a set of training samples. For an n-dimensional data, $n + 1$ linear equations or samples to solve the $n + 1$ weights are needed. Suppose $d_i = 1$ (or $d_i > 0$) represents class 1 and $d_i = -1$ (or $d_i < 0$) represents class 2, and the $n + 1$ training samples are given as $(\mathbf{x}_i, d_i)$, where

$$\mathbf{x}_i = (x_{i1}, x_{i2}, \ldots, x_{in}), \quad i = 1, 2, \ldots, n+1$$

Then, by substituting (8.2) with each of the training data $\mathbf{x}_i$, the set of weights w_i ($i = 0, 1, \ldots, n$) can be solved using the following $n + 1$ linear equations:

$$\sum_{j=0}^{n} w_j x_{ij} = d_i, \quad i = 1, 2, \ldots, n+1 \qquad (8.3)$$

Equation (8.3) can be written in matrix form

$$\begin{bmatrix} 1, & x_{11}, & x_{12}, \ldots, x_{1n} \\ 1, & x_{21}, & x_{22}, \ldots, x_{2n} \\ & & \cdots\cdots \\ 1, & x_{n1}, & x_{n2}, \ldots, x_{nn} \\ 1, & x_{n+1,1}, & x_{n+1,2}, \ldots, x_{n+1,n} \end{bmatrix} \begin{bmatrix} w_0 \\ w_1 \\ w_2 \\ \cdots \\ w_n \end{bmatrix} = \begin{bmatrix} d_1 \\ d_2 \\ \cdots \\ d_n \\ d_{n+1} \end{bmatrix} \qquad (8.4)$$

which in turn can be written as

$$\mathbf{Xw}^t = \mathbf{d}^T$$

where $\mathbf{w}^T$ and $\mathbf{d}^T$ are the transposes of $\mathbf{w}$ and $\mathbf{d}$, the solution of $\mathbf{w}$ is then given as (8.5)

$$\mathbf{w}^T = \mathbf{X}^{-1}\mathbf{d}^T \qquad (8.5)$$

8.1.2 An Optimal Solution

The above theoretical solution is just based on $n + 1$ or part of the training samples, therefore, it is not optimal. An optimal solution is usually given by minimizing the squared errors of $f(\mathbf{x})$ on the entire training data set of N data $(\mathbf{x}_i, d_i)$ and

$$\mathbf{x}_i = (x_{i1}, x_{i2}, \ldots, x_{in}), \quad i = 1, 2, \ldots, N$$

That is, to minimize the following total squared error:

$$E = \sum_{i=0}^{N-1} (f(\mathbf{x}_i) - d_i)^2$$
$$= \sum_{i=0}^{N-1} \left(\sum_{j=0}^{n} w_j x_{ij} - d_i \right)^2 \qquad (8.6)$$

By taking the partial derivative of E on w_k ($k = 0, 1, 2, \ldots, n$) and letting the partial derivative to be 0: $\frac{\partial E}{\partial w_k} = 0$, the following $n + 1$ linear equations are obtained:

$$\sum_{i=0}^{N-1} x_{ik} \left(\sum_{j=0}^{n} w_j x_{ij} - d_i \right) = 0, \quad k = 0, 1, 2, \ldots, n \qquad (8.7)$$

which is equivalent to the following:

$$\sum_{j=0}^{n} w_j \left(\sum_{i=0}^{N-1} x_{ik} x_{ij} \right) = \sum_{i=0}^{N-1} x_{ik} d_i, \quad k = 0, 1, 2, \ldots, n \qquad (8.8)$$

By solving the above $n + 1$ linear equations using the same method as solving (8.3), a set of weights $(w_0, w_1, w_2 \ldots w_n)$ is obtained which results is an optimal hyperplane. It can be observed that compared with (8.3), in an optimal solution algorithm, x_{ij} is replaced with $\sum_{i=0}^{N-1} x_{ik} x_{ij}$, while d_i is replaced with $\sum_{i=0}^{N-1} x_{ik} d_i$.

8.1.3 A Suboptimal Solution

Although the solution from (8.8) results in a more optimal decision boundary than that from (8.5), the solution of (8.8) would involve processing very large matrices which is computationally expensive and undesirable. This is because multimedia data usually has very high-dimensional features. An alternative approach is to use an iterative optimization algorithm to find a suboptimal solution to (8.2). Common practice is to use an error-driven weight-adaption technique which is basically a trial-and-error technique. The iterative optimization procedure is given in the following:

1. Initialize the weights $w_0, w_1, w_2 \ldots w_n$ with some small random values
2. Take the next training sample $\{\mathbf{x}, d\} = \{(x_1, x_2, \ldots, x_n), d\}$, $d = 1$ or -1
3. Compute $f(\mathbf{x}) = w_0 + w_1 x_1 + w_2 x_2 + \cdots + w_n x_n$
4. If $f(\mathbf{x}) \neq d$ (a misclassification), update $w_0 \leftarrow w_0 + cdk$ and $w_j \leftarrow w_j + cdx_j$, $j = 1, 2, \ldots, m$; where k and c are both positive constants
5. Repeat Steps 2–4 on each of the remaining training samples, until all the samples are correctly classified or the weights stop to change.

To demonstrate that the weights w_i or the hyperplane $f(\mathbf{x})$ are moving in the right directions, let f_{new} and f_{old} be the updated value and old value of $f(\mathbf{x})$ respectively. Because k, c and the feature value x_j are all positive, after the update in step 4, all the weights w_j become larger if $d = 1$ and all the weights w_j become smaller if $d = -1$. That means, if there is a misclassification, the decision function is updated according to the following rules:

$$\begin{cases} f_{new} > f_{old} & \text{if } d = 1 \\ f_{new} < f_{old} & \text{if } d = -1 \end{cases}$$

Therefore, in either case, the new hyperplane $f(\mathbf{x})$ is moving in the right direction with the updated weights, until the misclassified sample is located at the correct side of the hyperplane.

A linear classifier can only classify data which are linearly separable. However, this idea can be extended to build a nonlinear classifier. For example, we can convert a two-dimensional feature vector (x, y) in xy space to a five-dimensional feature vector $(u_1, u_2, u_3, u_4, u_5)$ in a higher dimensional space, where $u_1 = \sqrt{2}x$, $u_2 = \sqrt{2}y$, $u_3 = x^2$, $u_4 = \sqrt{2}xy$, $u_5 = y^2$. This would be the polynomial kernel $(1 + x + y)^2$. This is the key idea behind the kernel method which will be discussed in Sect. 8.3.

8.2 K-Nearest Neighbors Classification

K-nearest neighbors or K-NN is a simple algorithm that stores all available cases and classifies new cases based on a decision function (e.g., a distance measure).

Given a training dataset D and a distance measure *dist*:

- $(\mathbf{x}_i, y_i)$, $i = 1, 2, ..., N$
- $\mathbf{x}_i$ is a training data in R^n
- y_i is the corresponding class of the data $\mathbf{x}_i$, and $y_i \in \{c_j, j = 1, 2, ..., M\}$
- $dist\,(\mathbf{x} - \mathbf{x}_i) = \|\mathbf{x} - \mathbf{x}_i\|$.

A new observation data $\mathbf{x}$ is classified into one of the classes y_j using the following algorithm:

1. Input the new data $\mathbf{x}$
2. Compute the distance of $\mathbf{x}$ to all the training samples $\mathbf{x}_i$ in the dataset: $dist(\mathbf{x} - \mathbf{x}_i)$
3. Sort $dist(\mathbf{x} - \mathbf{x}_i)$ ($i = 1, 2, ..., N$) in ascending order and rank all the $\mathbf{x}_i$ accordingly: $\mathbf{x}_{r1}, \mathbf{x}_{r2}, ..., \mathbf{x}_{rk}, ..., \mathbf{x}_{rN}$
4. For the nearest neighbor (NN) classification, classify $\mathbf{x}$ to y_{r1}

   a. For a K-NN classification, classify $\mathbf{x}$ to the majority class y_{rp} among the top k ranked data: $\{\mathbf{x}_{r1}, \mathbf{x}_{r2}, ..., \mathbf{x}_{rk}\}$.

Although Euclidean (L_2) and city block distance (L_1) are a typical choice for the distance measure, any other distance can be used depending on the applications. The nearest neighbor (NN or 1-NN) results in too many classes, while K-NN gives more reliable classification results. This is because the values of k have a smoothing effect that makes the classifier more resistant to outliers. However, the performance of a K-NN classifier depends on the choice of k which is usually determined empirically.

Figure 8.1 demonstrates the comparison between an NN classifier and a K-NN classifier [1]. It can be seen from the two classification results, in the case of a NN classifier (after merging), outlier data points create small islands within a class (e.g., red point within the green class) and sharp corners on the class boundaries, those islands, and sharp corners likely lead to incorrect predictions; while the 5-NN

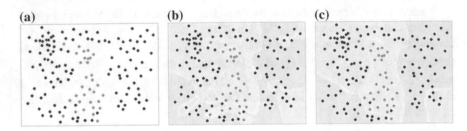

Fig. 8.1 Comparison between NN and K-NN. **a** The data to be classified; **b** classification result from a 1-NN classifier; **c** classification result from a 5-NN classifier

classifier smooths over these outliers, which lead to better classification on the data. However, the 5-NN classifier also causes misclassifications which are characterized by the blue dots in red region and red dots in green region. There can also be confusions by the tied votes among the five nearest neighbors (e.g., two neighbors are red, the next two neighbors are blue, and the last neighbor is green).

This kind of misclassification can be overcome to a certain extent by using the *weighted* K-NN. The idea is to give more weight to the neighbors with shorter distance to the test data than to the faraway neighbors. The commonly used weighted K-NN is the Gaussian weighted K-NN.

Unlike any other classifiers which are independent of the original training data once trained, a K-NN classifier is memoryless. If we analog a classifier to a connoisseur traveling around the world to judge (classify) different kinds of antiques for people. While other types of connoisseurs just need to take a toolkit summarizing the key characteristics of the antiques, a K-NN connoisseur will have to carry every kind of real antiques in his/her collection in order to make a new judgement. This may sound too cumbersome, however, one of the key advantages of a K-NN classifier is that it can classify data which are nonlinearly separable. This is the key idea behind the kernel-based support vector machine (SVM).

8.3 Support Vector Machine

In the previous two sections, we have introduced the linear classifier and K-NN classifier, both are key to understand SVM.

A linear classifier is simple and once trained, it is like a tool or a machine which can be used to tell if a data belongs to one of the classes. However, the disadvantages of a linear classifier include

- The solution is either not optimal or computationally expensive
- It cannot classify data which are nonlinearly separable

A K-NN classifier is also simple and can separate data nonlinearly. However, the disadvantages of a K-NN classifier include

- It is difficult to choose a k
- Dependence on training data.

Now that we have understood how the linear classifier and the K-NN classifier behave, we would like to build a classifier which takes advantage of both and overcomes their disadvantages. This is SVM.

An SVM is basically a binary linear classifier, however, with two prominent goals to achieve:

- To maximize the margin which separates the two classes (optimal)
- To use only a few training data (or support vectors) to determine the hyperplane which separates the two classes (efficient)

A kernel-based SVM adds another goal to

- Be able to classify data which are nonlinearly separable

As can be seen, once the three goals are achieve, we would truly build a machine which combines the advantages of both the linear classifier and the K-NN classifier, while overcomes their disadvantages.

To formulate SVM, we will start with the simple perceptron and the primal form of SVM. In the next, the dual form of SVM is introduced, and finally, the kernel based SVM is described in details.

8.3.1   The Perceptron

A perceptron is a binary linear classifier which is one of the simplest classifiers. Given an unknown data, the perceptron simply generates a linear prediction. The training process is the same as the linear classifier introduced in Sect. 8.1. The only difference is that a perceptron can do online learning, which means it can process the training data one at a time instead of having to taking the entire training dataset. Although it is simple, the perceptron is the key to understand both SVM and Artificial Neural Network (ANN) later on.

Given a training dataset D:

- $D = \{(\mathbf{x}_i, y_i), i = 1, 2, \ldots, N\}$
- $\mathbf{x}_i$ is a feature vector in n dimensional space: $\mathbf{x}_i = (x_{i1}, x_{i2}, \ldots, x_{in})$
- y_i is the corresponding class of the data $\mathbf{x}_i$, and $y_i \in \{-1, 1\}$
- $\langle \mathbf{x}_i, \mathbf{x}_j \rangle = \mathbf{x}_i \cdot \mathbf{x}_j$ is the dot product between two vectors

A perceptron is a binary linear classifier which is formulated as follows:

1. $f(\mathbf{x}) = \langle \mathbf{w}, \mathbf{x} \rangle + b$
2. Let $w_0 = b$ and $x_0 = 1$, then the above can be simply written as $f(\mathbf{x}) = \langle \mathbf{w}, \mathbf{x} \rangle$
3. $h(\mathbf{x}) = sign\ (f(\mathbf{x})) = y_i\ (f(\mathbf{x}))$
4. Take the next training data $(\mathbf{x}_i, y_i) \in D$
5. *if* $h(\mathbf{x}_i) \geq 0$, $\mathbf{w}_{k+1} \leftarrow \mathbf{w}_k$
6. *if* $h(\mathbf{x}_i) < 0$
    then $\mathbf{w}_{k+1} \leftarrow \mathbf{w}_k + \eta\ y_i\ \mathbf{x}_i,\ \eta > 0$
7. *Repeat from* 4

8.3.2   SVM—The Primal Form

8.3.2.1   The Margin Between Two Classes

Continue from the perceptron discussion and its training data assumption.

The perceptron gives us a hyperplane to separate the two classes of data, however, there are an infinite number of hyperplanes between two classes of data as shown in Fig. 8.2. The one resulted from the perceptron is just one of them, and it is nothing optimal. Although an optimal hyperplane was given in Sect. 8.1, it is optimal only in terms of minimizing the total error, and it is still far from the optimal hyperplane we *perceive*.

The optimal or the best hyperplane we *perceive* is the one separating the two classes with the maximal margin as shown in Fig. 8.3. That is the hyperplane we are going to find out.

Assume the two subspaces corresponding to the two classes of data are, respectively,

$$\langle \mathbf{w}, \mathbf{x} \rangle + b \geq 1\, for\, y_i = +1$$
$$\langle \mathbf{w}, \mathbf{x} \rangle + b \leq -1\, for\, y_i = -1$$

Fig. 8.2 Hyperplanes
between two classes of data

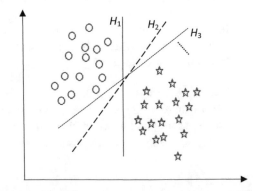

Fig. 8.3 The optimal
hyperplane H between two
classes of data

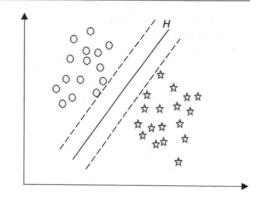

The above two inequalities can be combined into one

$$y_i(\langle \mathbf{w}, \mathbf{x} \rangle + b) - 1 \geq 0 \qquad (8.9)$$

The boundaries between the two subspaces are hyperplanes H_1 and H_2, respectively,

$$H_1: \quad \langle \mathbf{w}, \mathbf{x} \rangle + b - 1 = 0 \qquad (8.10)$$

$$H_2: \quad \langle \mathbf{w}, \mathbf{x} \rangle + b + 1 = 0 \qquad (8.11)$$

and the hyperplane between H_1 and H_2 is given as H_0:

$$H_0: \quad \langle \mathbf{w}, \mathbf{x} \rangle + b - 0 \qquad (8.12)$$

The two classes of data and the three hyperplanes separating them are shown in Fig. 8.4.

Our purpose is not only to find H_0 but also to maximize the distance between H_1 and H_2, which is the margin between the two classes of data. How to work out the distance between H_1 and H_2? Here is how it works out.

- Remember in a 2D space, a hyperplane is just a line which is expressed as: $ax + by + c = 0$. The constant c is called the intercept, and $|c|$ is associated with the distance from the origin to the line.
- This is also true in higher dimensional space. For example, in a 3D space, a plane is given as $Ax + By + Cz + D = 0$, and $|D|$ is associated with the distance from the origin to the plane. So on so forth.
- Therefore, the distance between H_1 and H_2 is equal to the difference between the distance of each of them to the origin.

Fig. 8.4 Two classes of data
and the hyperplanes
separating them

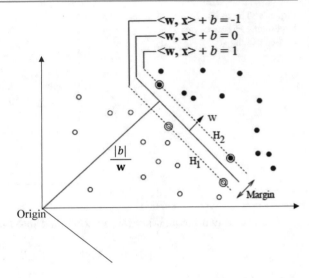

Specifically, based on the theory of geometry, the distance of a point (x_0, y_0, z_0) to a plane in 3D space: $Ax + By + Cz + D = 0$ is given as follows:

$$\frac{|Ax_0 + By_0 + Cz_0 + D|}{\sqrt{A^2 + B^2 + C^2}} \tag{8.13}$$

Because this is also true for higher dimensional space, accordingly, the distance from H_0 to the origin $(0, 0, \ldots, 0)$ in n-dimensional space is given as

$$\frac{|b|}{||\mathbf{w}||} \tag{8.14}$$

where $||\mathbf{w}||$ is the magnitude or length of vector $\mathbf{w}$, and the distance from H_1 and H_2 to the origin $(0, 0, \ldots, 0)$ in n-dimensional space is given by the following two respectively:

$$\frac{|b+1|}{||\mathbf{w}||} \quad \text{and} \quad \frac{|b-1|}{||\mathbf{w}||} \tag{8.15}$$

Therefore, by calculating the difference between the two terms of (8.15), the margin between the two hyperplanes H_1 and H_2 is obtained as

$$\frac{2}{||\mathbf{w}||} \tag{8.16}$$

The data points which lie on H_1 and H_2 are called *support vectors* (marked by circles in Fig. 8.3), which are both necessary and sufficient to define the boundary hyperplanes.

8.3.2.2 Margin Maximization

Therefore, based on (8.16), to maximize the distance between the two subspaces is equivalent to the following optimization problem:

$$Minimize: \quad f(\mathbf{w}) = \frac{1}{2}\|\mathbf{w}\|^2 = \frac{1}{2}\langle \mathbf{w}, \mathbf{w} \rangle \tag{8.17}$$

$$Subject\ to: \quad g_i(\mathbf{w}, b) = y_i(\langle \mathbf{w}, \mathbf{x}_i \rangle + b) - 1 \geq 0,\ i = 1, 2, \ldots, N \tag{8.18}$$

The above is a constrained optimization problem, and there are a few important facts to be pointed out [2]:

- b is one of the weights to be found because if we let $x_0 = 1$, then $w_0 = b$
- Equation (8.17) is a paraboloid in n-dimensional space
- A paraboloid has a single global minimum at the bottom
- Equation (8.18) is a hyperplane in n-dimensional space
- The solution to this constrained optimization problem is at the tangent point of the paraboloid and the hyperplane
- At the tangent point, the normal vectors or gradient vectors of both the paraboloid and the hyperplane are parallel
- That is, $\nabla f = \alpha_i \nabla g_i$, $(i = 1, 2, \ldots, N)$, where ∇ is the gradient and α_i is a constant.

Based on the above analysis, the optimization problem of (8.17) and (8.18) is equivalent to combining them into the following *Lagrange function* and solve $\nabla L = 0$ or $\partial_{\mathbf{w},\,b} L = 0$:

$$L(\mathbf{w},\ b,\ \alpha_i) = f(\mathbf{w}) - \sum_i \alpha_i g_i(\mathbf{w}, b) \tag{8.19}$$

$$
\begin{aligned}
L(\mathbf{w}, b, \alpha_i) &= \frac{1}{2}|\mathbf{w}|^2 - \sum_i \alpha_i [y_i(\langle \mathbf{w}, \mathbf{x}_i \rangle + b) - 1] \\
&= \frac{1}{2}\mathbf{w} \cdot \mathbf{w} - \sum_i \alpha_i y_i(\mathbf{w} \cdot \mathbf{x}_i + b) + \sum_i \alpha_i
\end{aligned}
\tag{8.20}
$$

8.3.2.3 The Primal Form of SVM

From (8.20), we can obtain the *primal form* of the SVM:

$$
\begin{aligned}
&Minimize: \quad L(\mathbf{w}, b, \alpha_i) = \frac{1}{2}\mathbf{w} \cdot \mathbf{w} - \sum_i \alpha_i y_i(\mathbf{w} \cdot \mathbf{x}_i + b) + \sum_i \alpha_i \\
&Subject\ to: \quad \alpha_i \geq 0, i = 1, 2, \ldots, N
\end{aligned}
\tag{8.21}
$$

8.3.3   The Dual Form of SVM

Although the primal form (8.21) let us to find the weights **w** and a hyperplane which separates the two classes of data with the maximum margin, the optimization is too expensive. Because we have to optimize two sets of parameters at the same time: **w** and α_i, this is very undesirable. Next, we want to make it more efficient.

Since (8.21) is a quadratic function, based on mathematics, at the global minima of the quadratic function, the gradient or the partial derivatives of $L(\mathbf{w}, b, \alpha_i)$ are 0. Therefore, we have

$$\nabla L(\mathbf{w}, b, \alpha_i) = \nabla f(\mathbf{w}) - \nabla \left[\sum_i \alpha_i g_i(\mathbf{w}, b) \right] = 0 \qquad (8.22)$$

Now let

$$\frac{\partial L}{\partial \mathbf{w}} = 0 \quad and \quad \frac{\partial L}{b} = 0$$

This leads to

$$\mathbf{w} = \sum_i \alpha_i y_i \mathbf{x}_i \quad and \quad \sum_i \alpha_i y_i = 0 \qquad (8.23)$$

Substituting the primal form (8.21) with (8.23) leads to the following *dual form* of the SVM:

$$\begin{aligned} Maximize: \quad & L_D = \sum_i \alpha_i - \tfrac{1}{2} \sum_{ij} \alpha_i \alpha_j y_i y_j \langle \mathbf{x}_i, \mathbf{x}_j \rangle \\ Subject\,to: \quad & \alpha_i \geq 0 \quad and \quad \sum_i \alpha_i y_i = 0 \end{aligned} \qquad (8.24)$$

To see why it has changed from minimization in the primal form to maximization in the dual form, let us have a good look at (8.24). The value of L_D is determined by the following three cases [2]:

- If the two features $\mathbf{x}_i, \mathbf{x}_j$ are completely dissimilar ($\mathbf{x}_i, \mathbf{x}_j$ are from different classes and are very different), their dot product $\langle \mathbf{x}_i, \mathbf{x}_j \rangle = 0$, that means, *features from different classes are far away from the boundaries between two classes don't contribute to L_D*
- If the two features $\mathbf{x}_i, \mathbf{x}_j$ are completely alike and from the same class, $\langle \mathbf{x}_i, \mathbf{x}_j \rangle \approx 1$ and $y_i\,y_j = 1$. Therefore, $\alpha_i\,\alpha_j\,y_i\,y_j\,\langle \mathbf{x}_i, \mathbf{x}_j \rangle$ would be positive and this would decrease the L_D. That means, *L_D downgrades similar features in the same class but far away from the boundaries between two classes*
- If the two features $\mathbf{x}_i, \mathbf{x}_j$ are completely alike but from the different class, $\langle \mathbf{x}_i, \mathbf{x}_j \rangle \approx 1$ but $y_i\,y_j = -1$. Therefore, $\alpha_i\,\alpha_j\,y_i\,y_j\,\langle \mathbf{x}_i, \mathbf{x}_j \rangle$ would be negative, this would increase L_D or maximize it. That means, *L_D is maximized with similar*

features from different classes or L_D *is maximized with features on the opposite boundaries of two classes*

To summarize the above analysis, by *maximizing* L_D, the dual form SVM

1. Emphasizes the feature vectors on the opposite boundaries between two classes
2. Ignores or suppresses those feature vectors far away from the boundaries between two classes.

This is exactly what we want because in terms of finding the hyperplanes separating the two classes with maximum margin, only those vectors on or close to the boundaries between the two classes matter most. Those feature vectors are called *support vectors* and the classifier defined by support vectors is called a *support vector machine*.

8.3.3.1 The Dual Form Perceptron

Because L_D is determined by the small number of support vectors on the boundaries between the two classes, not surprisingly, most of the α_i would be zero. Once α_i, $i = 1, 2, \ldots, N$ are solved, the weights for the hyperplane separating the two classes of data with the maximum margin are given as follows:

$$\mathbf{w} = \sum_i \alpha_i y_i \mathbf{x}_i \tag{8.25}$$

Therefore, the weight of the SVM hyperplane is just a linear combination of the training data, and this is consistent with the weight updating methods used in the linear and perceptron classifiers introduced earlier.

A set of α_i can be estimated using the *dual form* perceptron:

1. $f(\mathbf{x}) = \langle \mathbf{w}, \mathbf{x} \rangle + b$

   $= \sum_i \alpha_i y_i \langle \mathbf{x}_i, \mathbf{x} \rangle + b$
2. Take the next training data $(\mathbf{x}_j, y_j) \in D$
3. *if* $y_j \left(\sum_i \alpha_i y_i \langle \mathbf{x}_i, \mathbf{x}_j \rangle + b \right) \geq 0$
   then $\alpha_{i+1} \leftarrow \alpha_i$
4. *if* $y_j \left(\sum_i \alpha_i y_i \langle \mathbf{x}_i, \mathbf{x}_j \rangle + b \right) < 0$
   then $\alpha_{i+1} \leftarrow \alpha_i + \eta, \eta > 0$
5. *Repeat from step 2*

8.3.4 Kernel-Based SVM

8.3.4.1 The Dual Form SVM Versus NN Classifier

With the dual form SVM (8.24), we have successfully reduced the primal form optimization problem to optimizing just one set of parameters: α_i, $i = 1, 2, ...,$ N. This is much more efficient than (8.21). However, this is just a small part of the story about SVM, the more important part of the story is the transform of SVM optimization from testing $\langle \mathbf{w}, \mathbf{x}_i \rangle$ to testing $\langle \mathbf{x}_i, \mathbf{x}_j \rangle$. This is explained in the following:

- An n-dimensional data $\mathbf{x}$ is a feature vector in space, and geometrically, the dot product is defined as follows:

$$\langle \mathbf{x}_i, \mathbf{x}_j \rangle = \|\mathbf{x}_i\| \|\mathbf{x}_j\| \cos \theta \qquad (8.26)$$

  where θ is the angle between the two feature vectors $\mathbf{x}_i$ and $\mathbf{x}_j$
- In practice, the magnitudes of all feature vectors are normalized to unit or 1 so that they can be fairly matched
- Therefore, the dot production of two feature vector is just $\cos\theta$
- Because all feature values are positive, θ is between 0° and 90°
- For two feature vectors at the same direction or $\theta = 0^{\circ}$ (identical), the dot product is 1: $\cos\theta = 1$
- For two feature vectors at vertical angle or $\theta = 90^{\circ}$ (completely different), the dot product is 0: $\cos\theta = 0$
- For two feature vectors at an angle $0^{\circ} < \theta < 90^{\circ}$ (between similar to different), the dot product is between $(0, 1)$: $0 < \cos\theta < 1$
- Therefore, the dot product $\cos\theta$ actually measures the similarity between the two feature vectors, or, *the dot product is just the cosine distance between the two feature vectors*

Equipped with this key finding, now let us go back to (8.24):

$$\text{Maximize} \quad L_D = \sum_i \alpha_i - \frac{1}{2} \sum_{ij} \alpha_i \alpha_j y_i y_j \langle \mathbf{x}_i, \mathbf{x}_j \rangle$$

It is equivalent to

$$\text{Minimize} \quad \sum_{ij} \alpha_i \alpha_j y_i y_j \langle \mathbf{x}_i, \mathbf{x}_j \rangle \qquad (8.27)$$

Because $\langle \mathbf{x}_i, \mathbf{x}_j \rangle$ is just the distance between $\mathbf{x}_i$ and $\mathbf{x}_j$, by recalling what has been discussed in the K-NN section, we can see that (8.27) is just the *weighted nearest neighbors* classification.

Now, if we look at the dual form perceptron at the end of Sect. 8.3.3, the connection between SVM and K-NN is even clearer. The dual form classifier is given as (8.28)

$$f(\mathbf{x}) = \sum_i \alpha_i y_i \langle \mathbf{x}_i, \mathbf{x} \rangle + b \tag{8.28}$$

The classification of each training data $\mathbf{x}_j$ is done by testing

$$y_i \left(\sum_i \alpha_i y_i \langle \mathbf{x}_i, \mathbf{x}_j \rangle + b \right) \geq 0, \ j = 1, 2, \ldots, N \tag{8.29}$$

Again, this is just a *weighted nearest neighbors classifier.*

This is a significant development, because, by using the dual form, we not only make the SVM more efficient but also make it a *nonlinear classifier.*

8.3.4.2 Kernel Definition

These are some of the key points obtained from the above:

- The dot product is a kind of distance
- The dual form SVM is a kind of weighted nearest neighbors classifier
- The weighted nearest neighbors classifier is a nonlinear classifier

Now that we understand how an important role the dot product plays in the dual form SVM, we can extend this idea to any function behaves like a dot product.

It turns out the dot product of data points can be generalized as *kernelling.* Any function $K(\mathbf{x})$ which has the following property can be regarded as a *kernel*

$$K(\mathbf{x}_1, \mathbf{x}_2) = \langle \Phi(\mathbf{x}_1), \Phi(\mathbf{x}_2) \rangle \tag{8.30}$$

where $\Phi(\mathbf{x})$ is a function transforming feature vector $\mathbf{x}$ in one space R^m to another higher dimensional space R^n $(n > m)$. From the definition, a kernel behaves like a dot product, it takes two feature vectors as input and maps the two vectors to a scalar or a real value. The difference of a kernel from a dot product is that a kernel do the dot product at a higher dimensional space, called the Hilbert space. We will explain the benefit of doing this.

Not surprisingly, with this definition, the dot product itself is a kernel because

$$K(\mathbf{x}, \mathbf{y}) = \langle \mathbf{x}, \mathbf{y} \rangle = \langle \Phi(\mathbf{x}), \Phi(\mathbf{y}) \rangle \tag{8.31}$$

where $\Phi(\mathbf{x}) = \mathbf{x}$.

Given a kernel, the kernel-based SVM can now be written as

$$f(\mathbf{x}) = \sum_i \alpha_i y_i K \langle \mathbf{x}_i, \mathbf{x} \rangle + b \tag{8.32}$$

The questions now are:

1. Are there any other kernel functions than the dot product?
2. How useful is a kernel?

The answer to the first question is yes, there are many such kinds of kernel functions. Common kernel functions used in multimedia data classification include the following:

1. Quadratic Kernel

$$K(\mathbf{x}, \mathbf{y}) = \langle \mathbf{x}, \mathbf{y} \rangle^2 \text{ and } [1 + \langle \mathbf{x}, \mathbf{y} \rangle]^2 \tag{8.33}$$

2. Polynomial Kernel

$$K(\mathbf{x}, \mathbf{z}) = <\mathbf{x}, \mathbf{y}>^d \quad \text{and} \quad [1 + <\mathbf{x}, \mathbf{y}>]^d, d > 2 \tag{8.34}$$

3. Radial Basis Function (RBF) Kernel

$$K(\mathbf{x}, \mathbf{y}) = e^{-\gamma \|\mathbf{x}-\mathbf{y}\|^2}, \gamma > 0 \tag{8.35}$$

To demonstrate these functions having the kernel property of (8.30), let us assume $\mathbf{x}$ and $\mathbf{y}$ are in R^2 and $\mathbf{x} = (x_1, x_2)$, $\mathbf{y} = (y_1, y_2)$.

For $\langle \mathbf{x}, \mathbf{y} \rangle^2$:

$$\begin{aligned}
\langle \mathbf{x}, \mathbf{y} \rangle^2 &= [(x_1, x_2) \cdot (v_1, y_2)]^2 = (x_1 y_1 + x_2 v_2)^2 \\
&= x_1^2 y_1^2 + x_2^2 y_2^2 + 2 x_1 x_2 y_1 y_2 \\
&= \langle (x_1^2, \sqrt{2} x_1 x_2, x_2^2), (y_1^2, \sqrt{2} y_1 y_2, y_2^2) \rangle
\end{aligned} \tag{8.36}$$

$$\text{or} \quad \begin{aligned}
&= \langle (x_1^2, x_1 x_2, x_2 x_1, x_2^2), (y_1^2, y_1 y_2, y_2 y_1, y_2^2) \rangle \\
&= \langle \Phi(\mathbf{x}), \Phi(\mathbf{y}) \rangle
\end{aligned} \tag{8.37}$$

where $\Phi(\mathbf{x}) = (x_1^2, \sqrt{2} x_1 x_2, x_2^2)$ or $(x_1^2, x_1 x_2, x_2 x_1, x_2^2)$ is a function which maps a 2D feature vector to a 3D or 4D feature vector. Therefore, $\langle \mathbf{x}, \mathbf{y} \rangle^2$ is a kernel, and so is $\langle \mathbf{x}, \mathbf{y} \rangle^d$ when $d > 2$.

For $(1 + \mathbf{x} \cdot \mathbf{y})^2$, we have:

$$
\begin{aligned}
(1 + \mathbf{x} \cdot \mathbf{y})^2 &= [1 + (x_1, x_2) \cdot (y_1, y_2)]^2 \\
&= (1 + x_1 y_1 + x_2 y_2)^2 \\
&= 1 + 2x_1 y_1 + 2x_2 y_2 + 2x_1 x_2 y_1 y_2 + x_1^2 y_1^2 + x_2^2 y_2^2 \\
&= \langle (1, \sqrt{2}x_1, \sqrt{2}x_2, x_1^2, \sqrt{2}x_1 x_2, x_2^2), (1, \sqrt{2}y_1, \sqrt{2}y_2, y_1^2, \sqrt{2}y_1 y_2, y_2^2) \rangle \\
&= \langle \Phi(\mathbf{x}), \Phi(\mathbf{y}) \rangle
\end{aligned}
$$

$$(8.38)$$

where $\Phi(\mathbf{x}) = (1, \sqrt{2}x_1, \sqrt{2}x_2, x_1^2, \sqrt{2}x_1 x_2, x_2^2)$ is a function mapping a 2D feature vector to a 6 dimensional feature vector. Therefore, $(1 + \mathbf{x} \cdot \mathbf{y})^2$ is also a kernel, so is $(1 + \mathbf{x} \cdot \mathbf{y})^d$ for $d > 2$.

In general, a quadratic kernel $\langle \mathbf{x}, \mathbf{y} \rangle^2$ transforms an n dimensional vector $\mathbf{x} = (x_1, x_2, \ldots, x_n)$ to vector in $n(n + 1)/2$-dimensional space:

$$
\Phi : \mathbf{x} \longrightarrow (x_1^2, x_2^2, \ldots, x_n^2, x_1 x_2, x_1 x_3, \ldots, x_1 x_n, x_2 x_3, \ldots, x_2 x_n, \ldots, x_{n-1} x_n) \quad (8.39)
$$

For RBF $e^{-\gamma \|\mathbf{x} - \mathbf{z}\|^2}$, again assume $\mathbf{x}$ and $\mathbf{z}$ are in 2D, since

$$
\begin{aligned}
\|\mathbf{x} - \mathbf{z}\|^2 &= (x_1 - z_1)^2 + (x_2 - z^2)^2 \\
&= x_1^2 + z_1^2 - 2x_1 z_1 + x_2^2 + z_2^2 - 2x_2 z_2
\end{aligned}
$$

Without loss of generality, let $\gamma = \frac{1}{2}$, then we have

$$
\begin{aligned}
e^{-\gamma \|\mathbf{x} - \mathbf{z}\|^2} &= e^{-\frac{1}{2}\left(x_1^2 + z_1^2 - 2x_1 z_1 + x_2^2 + z_2^2 - 2x_2 z_2\right)} \\
&= e^{-\frac{1}{2}\left(x_1^2 + x_2^2\right)} e^{-\frac{1}{2}\left(z_1^2 + z_2^2\right)} e^{(x_1 z_1 + x_2 z_2)} \\
&= e^{-\frac{1}{2}\|\mathbf{x}\|^2} e^{-\frac{1}{2}\|\mathbf{z}\|^2} e^{\langle \mathbf{x}, \mathbf{z} \rangle} \\
&= C e^{\langle \mathbf{x}, \mathbf{z} \rangle} \\
&= C \sum_{n=0}^{\infty} \frac{\langle \mathbf{x}, \mathbf{z} \rangle^n}{n!} \quad \text{(Taylor expansion of } e^x\text{)} \\
&= C \sum_{n=0}^{\infty} \frac{K_{poly(n)}(\mathbf{x}, \mathbf{z})}{n!}
\end{aligned}
$$

$$(8.40)$$

where $C = e^{-\frac{1}{2}\|\mathbf{x}\|^2} e^{-\frac{1}{2}\|\mathbf{z}\|^2}$ is a constant because feature vector are normalized to unit length, and $K_{poly(n)}(\mathbf{x}, \mathbf{z})$ is a polynomial kernel [3]. Therefore, the RBF is a kernel because the sum of kernels is also kernel (see the following). Equation (8.40) shows that the RBF maps a vector into a space with infinite dimensions.

8.3.4.3 Building New Kernels

It can be shown that the following rules are true:

1. The sum of two kernels is also a kernel

$$K(\mathbf{x}, \mathbf{y}) = K_1(x, y) + K_2(x, y) \tag{8.41}$$

2. A scalar times a kernel is also a kernel

$$K(\mathbf{x}, \mathbf{y}) = aK_1(x, y) \tag{8.42}$$

3. The product of two kernels is also a kernel

$$K(\mathbf{x}, \mathbf{y}) = K_1(x, y) \times K_2(x, y) \tag{8.43}$$

Therefore, by using these rules and existing kernels, we may build more kernels for different applications.

8.3.4.4 The Kernel Trick

Now that we have defined the kernels and understood their behaviors, the next is to answer the *second question* we mentioned earlier. That is, why kernels or why we transform a feature vector to a higher dimensional space? It appears the dual form SVM is good enough because it not only gives us a SVM but also let us do nonlinear classification. So what is the benefit of using kernels?

There are two reasons to use a kernel instead of just the dot product.

- One is to transform nonlinear data in lower dimensional space to linear data in higher dimensional space so that they can be separated linearly using the SVM.
- The other is to have more and better choices of distance measurement than the dot product, so as to improve the performance of an SVM.

To demonstrate how a kernel can transform nonlinear data into linear data, we will use the quadratic kernels as examples.

Consider the following 1D binary data (red and green dots) which is a nonlinear data because it cannot be separated by a point or a line.

Now map each of the samples using the following function:

$$\Phi : x \rightarrow \{x, x^2\} \tag{8.44}$$

Φ is a quadratic mapping, it transforms a 1D line into a 2D parabola:

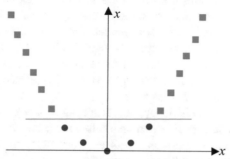

By transforming the 1D data into a 2D space, now the data in 2D space can be separated using a line (blue) or linearly separable. This is exactly the first reason for using kernel. This phenomenon can also be demonstrated using a 2D nonlinear data (Fig. 8.5) [4]. By using the following mapping function to map the 2D data on the left of Fig. 8.5 to a paraboloid in 3D space, the data can now be separated using a 2D plane and is linearly separable:

$$\Phi : (x_1, x_2) \rightarrow (x_1^2, \sqrt{2}x_1x_2, x_2^2) \tag{8.45}$$

Because the dot product is kind of distance measure, therefore, all kernels behave like a distance measure. Just like a good distance measure is crucial to a classifier, the choice of a good kernel can affect a classifier significantly. This is the reason why a kernel-based SVM is always better than an SVM just using the simple dot product.

(a)   **(b)**

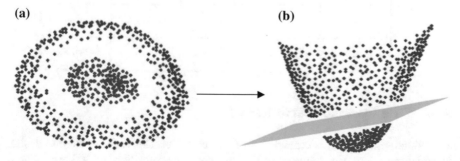

Fig. 8.5 Mapping of nonlinear data to linear data in higher dimensional space. **a** An original nonlinear data in 2D space; **b** transformed data in 3D space using a quadratic mapping function φ

Although the use of kernel gives us the advantage to do nonlinear classification, the explicit mapping from a lower dimensional space to a higher dimensional space is undesirable and can be expensive in terms of computation, given the fact that a feature vector usually has high dimension. Furthermore, data representation using very high dimension in Hilbert space is inefficient too.

Fortunately, the mapping does not need to be done explicitly. So long it is a kernel, it *implicitly* maps a data to one in another space which is linearly separable. Put the other way, a kernel is just a dot product (*implicit*) regardless the space where the dot product is done, and according to (8.32), a kernel SVM is just a weighted nearest neighbors classifier by which a data can always be separated nonlinearly. Therefore, all we need to do for a kernel SVM is just to replace the dot product with a kernel. This is called the *kernel trick*.

To further improve the efficiency, in practice, an $N \times N$ kernel (or Gram) matrix is precomputed for a dataset of N elements before the actual learning, so that there is no need to recompute the dot products at every iteration of the optimization. A kernel matrix K has the following properties:

- K is a positive definite matrix
- $K(i,j) = K(\mathbf{x}_i, \mathbf{x}_j) = \langle \Phi(\mathbf{x}_i), \Phi(\mathbf{x}_j) \rangle$ (implicit dot product in higher dimension space)
- K is symmetric, or $K(i, j) = K(j, i)$
- $K(i, j)$ measures the similarity between ith and jth training samples in feature space

$$
K = \begin{array}{|c|c|c|c|c|}
\hline
K(\mathbf{x}_1, \mathbf{x}_1) & K(\mathbf{x}_1, \mathbf{x}_2) & K(\mathbf{x}_1, \mathbf{x}_3) & \text{.......} & K(\mathbf{x}_1, \mathbf{x}_N) \\
\hline
K(\mathbf{x}_2, \mathbf{x}_1) & K(\mathbf{x}_2, \mathbf{x}_2) & K(\mathbf{x}_2, \mathbf{x}_3) & \text{.......} & K(\mathbf{x}_2, \mathbf{x}_N) \\
\hline
K(\mathbf{x}_3, \mathbf{x}_1) & K(\mathbf{x}_3, \mathbf{x}_2) & K(\mathbf{x}_3, \mathbf{x}_3) & \text{.......} & K(\mathbf{x}_3, \mathbf{x}_N) \\
\hline
\text{.......} & \text{.......} & \text{.......} & \text{.......} & \text{.......} \\
\hline
K(\mathbf{x}_N, \mathbf{x}_1) & K(\mathbf{x}_N, \mathbf{x}_2) & K(\mathbf{x}_N, \mathbf{x}_3) & \text{.......} & K(\mathbf{x}_N, \mathbf{x}_N) \\
\hline
\end{array}
$$

8.3.5 The Pyramid Match Kernel

A well-designed kernel is crucial to an SVM classifier. Conventional kernel design is independent of the feature itself. However, the selection of a kernel for a particular type of features is difficult because there is no natural connection between a feature and a kernel. Consequently, the selection of kernel for an SVM classifier is often arbitrary or empirical at best. The Pyramid Match Kernel or PMK [5] is a method to design a kernel which matches the specific type of image features.

The idea is to extract a pyramid histogram feature at different level of resolutions and build a kernel using a weighted sum of histogram intersections. The idea of the PMK is described in details in the following:

- Start with image X itself as level 0 and the total number of levels is L.
- Divide image into grids at different levels of resolutions. The grid at level l has a total of $2^l \times 2^l = 4^l$ cells, with 2^l cells along each dimension.
- A histogram is computed for each block at each level of resolutions.
- Histograms at each level l are given a different weight.
- The weighted histograms from all levels are concatenated as the pyramid histogram of the image.
- A kernel of weighted histogram intersection is built for the SVM.

$$K(X, Y) = \sum_{l=1}^{L} \alpha_l k^l(X_m, Y_m) \tag{8.46}$$

where X_m and Y_m are two weighted pyramid histograms and k is the histogram intersection.

- The idea is illustrated in Fig. 8.6 [5].

Let X^l and Y^l stand for the histograms of X and Y at level l, then the number of matches at this level is given by the histogram intersection

$$k^l = \sum_{i=1}^{4^l} \min[X^l(i), Y^l(i)] \tag{8.47}$$

where $l = 0, 1, 2, ..., L$. k^l at lower levels represent global features while k^l at higher levels represent local features. Since global features can cause more confusion than local features, global features should be given less weights than local features. Therefore, the weight given to level l is set to $1/2^{L-l}$, which is inversely proportional to the block width at that level. Since the total weights must sum to 1, the combined matching result between two images is given in (8.48).

$$K(X, Y) = \frac{1}{2^L} k^0 + \sum_{l=1}^{L} \frac{1}{2^{L-l+1}} k^l \tag{8.48}$$

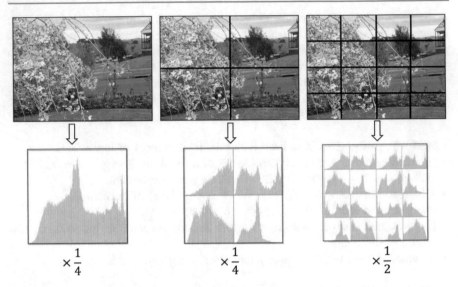

Fig. 8.6 Computation of pyramid match kernel. An image is divided into three levels of grids. At each level of the grid, a histogram is computed for each block of the grid. Histograms at each level are given a weight and the weighted histograms are then concatenated as a feature vector

The next is to prove (8.48) is a kernel. Because a linear combination of kernels is also a kernel, we just need to prove each histogram intersection k^l is a kernel.

Let X_m and Y_m be the histograms of two images or image blocks X and Y. Each image has N pixels. We can then represent X_m and Y_m as two $N \times m$ dimensional binary vectors [6].

$$X_m = (\overbrace{1, 1, \ldots, 1}^{x_1}, \underbrace{0, 0, \ldots, 0}_{N-x_1}; \overbrace{1, 1, \ldots, 1}^{x_2}, \underbrace{0, 0, \ldots, 0}_{N-x_2}; \ldots; \overbrace{1, 1, \ldots, 1}^{x_m}, \underbrace{0, 0, \ldots, 0}_{N-x_m}) \quad (8.49)$$

$$Y_m = (\overbrace{1, 1, \ldots, 1}^{y_1}, \underbrace{0, 0, \ldots, 0}_{N-y_1}; \overbrace{1, 1, \ldots, 1}^{y_2}, \underbrace{0, 0, \ldots, 0}_{N-y_2}; \ldots; \overbrace{1, 1, \ldots, 1}^{y_m}, \underbrace{0, 0, \ldots, 0}_{N-y_m}) \quad (8.50)$$

With the above representation, the histogram intersection of X_m and Y_m is given as the dot product of the two histograms:

$$k(X_m, Y_m) = X_m \cdot Y_m \quad (8.51)$$

Therefore, k^l is a kernel and as a result, $K(X, Y)$ in (8.48) is also a kernel.

Fig. 8.7 Histogram intersection of two normal distributions

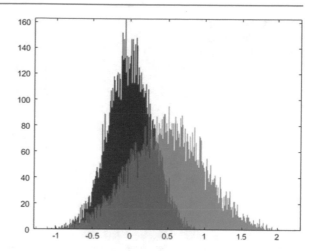

A histogram is a statistical feature, it captures the feature distribution in an image or an image block. Histogram intersection tells how much area two distributions share, the more area they share, the more similar the two distributions are. Figure 8.7 shows an example of histogram intersection. The shared region is about 33% of the two histograms, therefore, the similarity between the two histograms is about 33%.

8.3.6 Discussions

Kernel-based support vector machine is essentially a training-based nearest neighbor classifier. The use of dot product transforms the support vector machine into a nonlinear nearest neighbor classifier. Traditional nearest neighbor has two limitations, the determining of k and it does not support training. However, if the training set is sufficiently large, both limitations can be overcome. First, the k can be determined empirically. The second limitation can be overcome by determining the class boundary with a piecewise linear approximation. For example, the class boundary of the following data can be approximated by 5 hyperplanes, which can then be used to classify new data (Fig. 8.8).

Although the piecewise linear boundary given by the K-NN is not optimal as the boundary provided by the kernel-based SVM, in terms of classification, the effectiveness of the two classifiers can be comparable. However, it would not be as efficient as SVM.

Fig. 8.8 Approximation of
class boundary using
piecewise hyperplanes

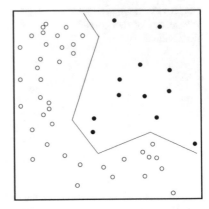

8.4 Fusion of SVMs

8.4.1 Fusion of Binary SVMs

An SVM is essentially a binary classifier. However, automatic image classification
and annotation needs a multi-class classifier. The most common approach is to train
a separate SVM for each concept c and each SVM generates a decision value $d_c(\mathbf{x})$.
During the testing phase, the decisions from all classifiers are fused to obtain the
final class label of a test image. Figure 8.9 demonstrates this two level fusion
process [7, 8]. The first level consists of multiple binary classifiers and the second
level fuses the decisions from the first level classifiers.

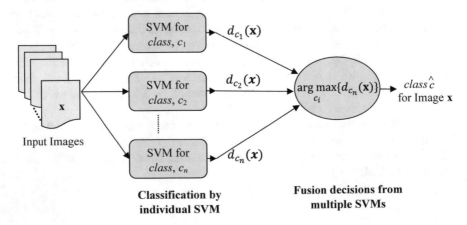

Fig. 8.9 A fusion of binary SVM classifiers

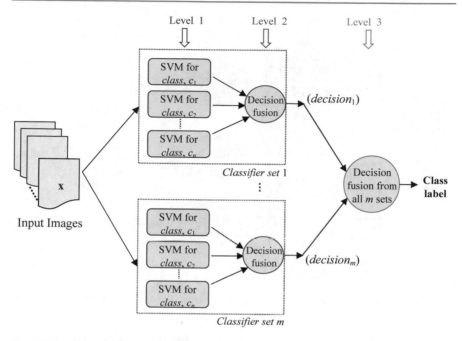

Fig. 8.10 A 3 levels fusion of SVMs

8.4.2 Multilevel Fusion of SVMs

The above approach can be regarded as a base level fusion, it works well for a small number of concepts. The quality of classification degrades with the increase of the number of concepts due to the increase of the noise and class imbalance in the training data. To be more robust, *multiple sets* of base level fusion of SVMs can be merged to make a more powerful fusion as shown in Fig. 8.10 [7, 8]. Each set of SVMs in level 1 and level 2 is similar to the base level fusion shown in 8.9 and independently classifies an input image, the final decision is fused from the decisions of all the individual sets at level 3.

The key advantage of using multiple sets of SVMs is to learn a more accurate and robust classifier using different types of SVMs, such as classification SVMs, regression SVMs, SVMs with/without soft margins, etc.

8.4.3 Fusion of SVMs with Different Features

Fusion of classifiers can also be done with combination of different types of features. For example, both global and local features can be used to train two different sets of SVMs at level 1 as shown in Fig. 8.11 [7, 8]. The results from the two sets of SVMs are then fused in two steps. First, decisions of each concept made by each set

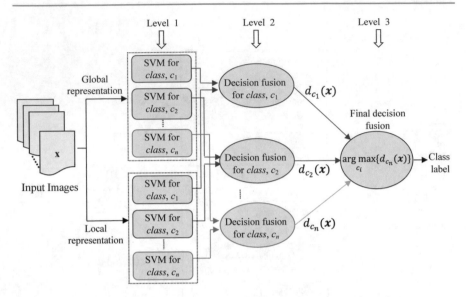

Fig. 8.11 A 3 levels class-by-class fusion of SVMs with both global and local features

of SVM are fused at level 2. Next, the final decision is made using a maximization at level 3.

Although the fusion methods discussed in this section are shown as fusion of SVMs, they can also be applied to fusion of different types of classifiers such as Bayesian, ANN, DT, etc.

8.5 Summary

SVM is basically a supervised linear classifier which divides a datasct into two classes with a hyperplane in data space. However, different from an ordinary linear classifier, it offers an optimal hyperplane which separates two classes of data with maximum margin between them. The data points make the hyperplane are called the support vectors. SVM works by repeat guessing with candidate hyperplanes until the optimal hyperplane is found.

The biggest progress of SVM is the kernel-based SVM which achieves nonlinearity without the use of networking like ANN. Nonlinearity of SVM is achieved through transforming data into higher space so that they can be separated linearly. Due to the kernel trick, this transformation is even unnecessary so long as the distance is a kernel. This makes SVM is very efficient compared with ANN.

However, SVM is essentially a binary classifier or non-probabilistic classifier. This makes it less robust than other probabilistic classifiers such as Bayesian classifiers and DT. In addition, a multi-class SVM needs to be achieved through fusion or assembly.

References

1. Karpathy A (2019) Stanford CS Class CS23. In: Convolutional neural networks for visual recognition. http://cs231n.github.io/classification/. Accessed Feb 2019
2. Berwick R (2019) An idiot's guide to support vector machines (SVMs). http://web.mit.edu/6.034/wwwbob/svm-notes-long-08.pdf. Accessed Feb 2019
3. Bernstein M (2019) The radial basis function kernel. http://pages.cs.wisc.edu/~matthewb/pages/notes/pdf/svms/RBFKernel.pdf. Accessed Feb 2019
4. Cambridge Spark (2019) Support vector machines. http://beta.cambridgespark.com/courses/jpm/05-module.html. Accessed Feb 2019
5. Lazebnik S, Schmid C, Ponce J (2006) Beyond bags of features: spatial pyramid matching for recognizing natural scene categories In: CVPR06, vol II, pp 2169–2178
6. Barla A, Odone F, Verri A (2003) Histogram intersection kernel for image classification. In: ICIP03
7. Islam M (2009) SIRBOT—semantic image retrieval based on object translation. PhD thesis, Monash University
8. Zhang D, Islam M, Lu G (2012) A review on automatic image annotation techniques. Pattern Recognit 45(1):346–362

Artificial Neural Network

9

Law of nature is The Way.

9.1 Introduction

When comes to learning and classifications, no other tool is more efficient and powerful than human brains. Therefore, there is sufficient motivation to design a machine learning tool which simulates human brains. This is further encouraged by the recent research findings on human brain from both cognitive science and biology. It is believed that a human brain is consisting of 10s of billions of neurons interconnected into a sophisticated network. The neurons in a brain are organized into functional units or regions, such as regions for visual, auditory, motion, reasoning, speech, etc. An individual neuron is shown in Fig. 9.1.

It is found that a neuron receives inputs from its dendrites, processes them in the cell body, and transmits the output signal to other neurons through its axon. The inputs the neuron received can be either *excitatory* or *inhibitory*. When there are more excitatory inputs than inhibitory inputs, the neuron is activated and a signal is transmitted out through the axon; otherwise, no signal is generated.

Then, neurons in many regions of a human brain are further organized into layers to create a layered network. Neurons from one layer usually receive inputs from neurons in an adjacent layer. Connections between layers are mostly in one direction, moving from low-level layer sensors like eyes or ears to higher coordination and reasoning layer [1].

With these understandings of human brains and neurons, it is possible to design artificial neurons and an artificial neural network or ANN.

© Springer Nature Switzerland AG 2019
D. Zhang, *Fundamentals of Image Data Mining*, Texts in Computer
Science, https://doi.org/10.1007/978-3-030-17989-2_9

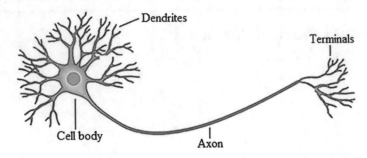

Fig. 9.1   A neuron of human brain

9.2   Artificial Neurons

The design of an artificial neural network starts with the modeling of artificial neurons. From the above understandings, a neuron is basically a unit which receives inputs and generates an output. Specifically, a biological neuron consists of three components: the inputs (dendrites), an activation or processing unit (cell body) and an output (axon). Electronically, these three components can be respectively represented as a set of inputs x_i, a weighted sum of the inputs Σ, and a threshold of the weighted sum. The alignment of an electronic neuron and a biological neuron is shown in Fig. 9.2.

The weighted sum and the thresholding are usually merged into a single activation unit and the axon is replaced with an output signal. Therefore, the simplified artificial neuron is shown in Fig. 9.3.

It turns out that an artificial neuron is just a binary linear classifier. Given an input $\mathbf{x} = (x_1, x_2, \ldots, x_n)$, a weighted sum Σ is calculated and compared with a threshold T:

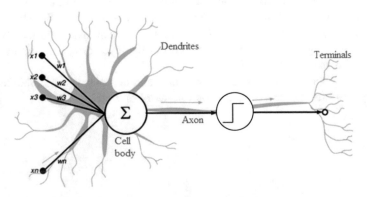

Fig. 9.2   The alignment of an artificial neuron with a biologic neuron

Fig. 9.3 The modeling of an artificial neuron

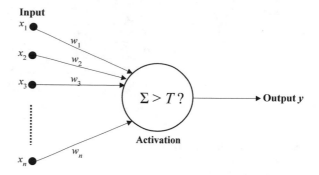

$$\Sigma = w_1 x_1 + w_2 x_2 + \cdots + w_n x_n > T \qquad (9.1)$$

If $\Sigma > T$, the neuron is activated and an output signal y is sent out. In other words, the activation of the neuron or output y is based on the following rule:

$$y = \begin{cases} 1 & \Sigma > T \\ 0 & \Sigma < T \end{cases} \qquad (9.2)$$

It is more convenient to combine both Σ and T and rewrite (9.1) as follows:

$$\begin{aligned} D = \Sigma - T &= w_1 x_1 + w_2 x_2 + \cdots + w_n x_n - T \\ &= -T + w_1 x_1 + w_2 x_2 + \cdots + w_n x_n \end{aligned} \qquad (9.3)$$

For notation purpose, let $w_0 x_0 = -T$ which represents a constant and $x_0 = 1$, then (9.3) becomes

$$D = w_0 x_0 + w_1 x_1 + w_2 x_2 + \cdots + w_n x_n \qquad (9.4)$$

And (9.2) becomes

$$y = \begin{cases} 1 & D > 0 \\ 0 & D < 0 \end{cases} \qquad (9.5)$$

We can use this artificial neuron to do many simple linear classifications. One of them is to simulate the logic gates, such as the AND, OR, NAND, NOR, etc.

9.2.1   An AND Neuron

Let us start with the AND gate, the AND function is given in Table 9.1.

Now, if we set the activation or threshold value as $T = 1.5$ (or any value between 1 and 2), only input (1, 1) will be activated, this is exactly what we want. Therefore, the classifier for the AND function is given as follows:

Table 9.1 AND gate

Input		Output	Sum
x_1	x_2		
0	0	0	0
0	1	0	1
1	0	0	1
1	1	1	2

Fig. 9.4 The data of AND can be separated by a single line $-1.5 + x_1 + x_2 = 0$

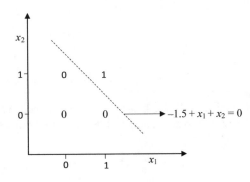

Fig. 9.5 The neuron which implements the AND function

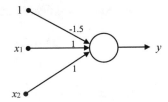

$$y = -1.5 + x_1 + x_2 \tag{9.6}$$

In space, the classifier of (9.6) is represented by the following hyperplane which is a line in this case:

$$-1.5 + x_1 + x_2 = 0 \tag{9.7}$$

The AND data and the linear classifier is shown in Fig. 9.4. The neuron which implements the AND function is shown in Fig. 9.5 [1].

9.2.2   An OR Neuron

Similarly, the classifier of an OR function and the neuron that implements the OR function are shown in Figs. 9.6 and 9.7, respectively.

An artificial neuron is called a node in an artificial neural network and is usually represented by a circle.

Fig. 9.6 The data of OR can be separated by a single line $-0.5 + x_1 + x_2 = 0$

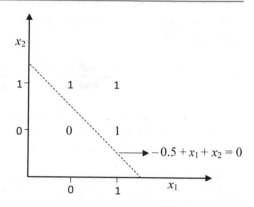

Fig. 9.7 The neuron which implements the OR function

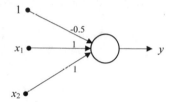

9.3 Perceptron

The neuron we designed above can do binary and linear classification; however, both the weights and the threshold are predetermined by a human designer. A biological neuron of a human brain, on the other hand, can learn from new instances and memorize. It would be desirable that an artificial neuron can also learn and memorize. This is done by feeding the neuron with a set of known data or pre-labeled data and learn the weights by using certain criterion or algorithm such as minimizing the total error.

From (9.4), the decision function of a neuron is given as follows:

$$Y = w_0 + w_1 x_1 + w_2 x_2 + \cdots + w_n x_n \tag{9.8}$$

To train the neuron:

- A training set $S = \{(\mathbf{x}_q, y_q), q = 1, 2, \ldots, N\}$ is collected,
- where y_q is the desired output of qth sample feature vector $\mathbf{x}_q = (x_{q1}, x_{q2}, \ldots, x_{qn})$.
- The training is then to minimize the squared error or MSE:

$$E = E(w_1, w_2, \ldots, w_n) = \frac{1}{2} \sum_{q=1}^{N} (Y_q - y_q)^2 \tag{9.9}$$

- where $Y_q = w_0 + w_1 x_{q1} + w_2 x_{q2} + \cdots + w_n x_{qn}$.

The minimization follows the *steepest descent* direction which is given by the gradient vector of

$$\left(-\frac{\partial E}{\partial w_0}, -\frac{\partial E}{\partial w_1}, -\frac{\partial E}{\partial w_2}, \ldots, -\frac{\partial E}{\partial w_n}\right) \tag{9.10}$$

If a sample is misclassified and the actual output Y is different from the correct output y, we would want to change the weights so that E is minimized. Therefore, the *steepest descent* algorithm to minimize E is given by the following algorithm:

1. Choose an initial weight set of w_0, w_1, w_2, ..., w_n and a positive constant c.
2. For $i = 0, 1, 2, ..., n$, compute the partial derivatives of $\partial E/\partial w_i$ and let $w_i = w_i - c(\partial E/\partial w_i)$.
3. Repeat step 2 until w_0, w_1, w_2, ..., w_n stop to change.

By combining (9.9) and (9.10), the partial derivatives $\partial E/\partial w_i$ in the above algorithm is given by

$$\frac{\partial E}{\partial w_i} = \left(Y_q - y_q\right)x_{qi}, \quad i = 0, 1, 2, \ldots, n \tag{9.11}$$

Therefore, the MSE learning algorithm of a *perceptron* is given as follows:

1. Choose an initial weight set of w_0, w_1, w_2, ..., w_n and a positive constant c.
2. For each of samples $q = 1, 2, ..., N$, compute $Y_q = \sum_{i=0}^{n} w_i x_{qi}$.
3. Let $w_i = w_i - c(Y_q - y_q)x_{qi}$ for $i = 0, 1, 2, ..., n$.
4. Repeat steps 2 and 3 until w_0, w_1, w_2, ..., w_n stop to change.

9.4 Nonlinear Neural Network

The perceptron is essentially a two-layer neural network with an input layer and an output layer, it can be trained to classify any linear data which can be separated by a hyperplane. However, for nonlinear data such as XOR (Fig. 9.8) and other data with convex data regions (Fig. 9.9), they cannot be separated by a single hyperplane in space, and consequently they cannot be classified by a perceptron.

Although a convex data region in space (a region is convex if any two data points can be connected by a line segment inside the region) cannot be separated by a single hyperplane, its boundary can be approximated by the intersection of a finite number of hyperplanes. For example, the XOR data in Fig. 9.8 can be separated by the following two half planes in 2D space:

$$-0.5 + x_1 + x_2 > 0 \quad \text{and} \quad -1.5 + x_1 + x_2 < 0 \tag{9.12}$$

Fig. 9.8 Outputs of an XOR cannot be separated by a single line in 2D space

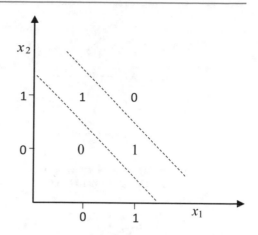

Fig. 9.9 The convex data region in the center cannot be separated by a single line in 2D space

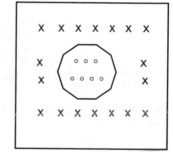

The two neurons represent the two half planes in (9.12) which are given by

$$-0.5 + x_1 + x_2 > 0 \quad \text{and} \quad 1.5 - x_1 - x_2 > 0 \tag{9.13}$$

Now, by AND(ing) (Fig. 9.5) the two neurons of (9.13), a neural network with a middle layer or hidden layer is created which can separate the XOR outputs. The neural network with hidden layer is shown in Fig. 9.10 [1].

This idea of three-layer neural network can be easily extended to classify any generic convex nonlinear data like the one shown in Fig. 9.9. All we need now is more nodes in the second layer or middle layer (Fig. 9.11). Usually, a small number of nodes in the middle layer are sufficient to separate a convex region; however, more middle layer nodes produce a smoother region boundary.

By extending the above idea of classifying *convex* nonlinear data using neural network with a hidden layer, it is also possible to classify *non-convex* data using neural network. This is because a *non-convex* region can always be approximated by the union of a finite number of convex regions.

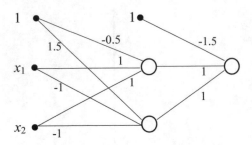

Fig. 9.10 A three-layer neural network to implement XOR. The two linear classifiers at the hidden layer are ANDed at the third layer

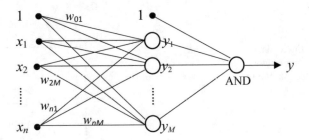

Fig. 9.11 A three-layer neural network with a hidden layer which can classify generic convex nonlinear data

Fig. 9.12 A non-convex data region (black dots) is approximated by three convex regions

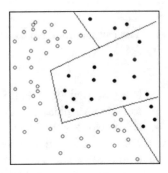

Therefore, it is possible to create a number of convex nodes in the third layer using the method discussed above and combine those convex nodes in the third layer using a logic OR in the fourth layer.

For example, the non-convex data region (black dots) in Fig. 9.12 can be approximated by three convex data regions, which can then be classified using a four-layer neural network shown in Fig. 9.13.

This indicates that any *nonlinear* data can be classified by a four-layer neural network.

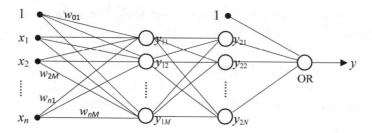

Fig. 9.13 A four-layer neural network which can classify any nonlinear data

9.5 Activation and Inhibition

It should be noted that the activation or threshold function is crucial to a neural network, because it is shown that any neural network without thresholding is equivalent to a two-layer network which cannot separate nonlinear data. However, the binary threshold function is not desirable because it is not continuous. This non-continuity can cause the network to take a very long time to converge or even not converge. This is because the gradients from the MSE are differential values and are small, so the changes in the weights are also small. As the result, the small changes to the weights are usually not enough to pass the threshold or generate an output signal.

9.5.1 Sigmoid Activation

It is desirable to have a continuous activation function which changes continuously from 0 to 1 instead of jumping from 0 to 1. Among the many proposed continuous activation functions, the most widely used is the *sigmoid function*. The S-shaped *sigmoid function* is defined as (9.14) and the shape of the function is shown in Fig. 9.14.

Fig. 9.14 Sigmoid function

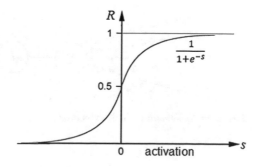

The $R(s)$ function guarantees an output signal while preserving the thresholding functionality of an activation function.

$$R(s) = \frac{1}{1 + e^{-s}} \tag{9.14}$$

The $R(s)$ function has the following important properties:

1. $\lim_{s \to -\infty} R(s) = 0$
2. $\lim_{s \to \infty} R(s) = 1$
3. $R(0) = \frac{1}{2}$
4. $\dfrac{dR}{ds} = R(1 - R)$ $\qquad\qquad\qquad\qquad\qquad\qquad\qquad\qquad$ (9.15)

This final property has a convenient use in the following backpropagation algorithm. $R(1 - R)$ is close to 0 at both ends of region (0, 1).

9.5.2 Shunting Inhibition

It is known that a biological neuron can be either *excitatory* or *inhibitory*. An inhibitory signal prevents impulse from arising in the receiving neuron. This inhibitory phenomenon can be represented mathematically as reducing the excitatory potential by division. This is called *shunting inhibitory* or SI. The idea is to learn a dual set of weights D_j and divided the original output of a neuron with the dual output adjusted by a decay factor. An SI neuron is illustrated in Fig. 9.15 [2].

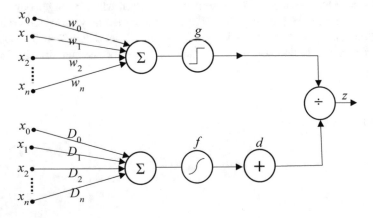

Fig. 9.15 A shunting inhibitory neuron

Mathematically, the output from an SI neuron is given by (9.16) [2].

$$z = \frac{g(\sum_{j=1}^{n} w_j x_j + b)}{d + f(\sum_{j=1}^{n} D_j x_j + a)} \tag{9.16}$$

where

- z is the output of the shunting neuron,
- x_j is the jth input,
- C_j and D_j are the connection weights of the jth input,
- $a = w_0 x_0$ and $b = D_0 x_0$ are biases,
- d is the passive decay rate,
- f and g are activation functions,
- n is the number of inputs from the previous layer,
- $d + f(\sum_{j=1}^{n} D_j x_j + a) > 0$.

An ANN designed with SI neurons is called an SIANN. The key to an SI neural network is to use different activation functions f and g in a layer so that only the strongest neurons are activated. Experiments show that when f and g are hyperbolic tangent function and exponential function, respectively, the network has better convergence.

9.6 The Backpropagation Neural Network

9.6.1 The BP Network and Error Function

One of humans' great learning skills is learning from mistakes or errors, this is because the mistakes/errors provide important feedback to improve the original learning process. Unfortunately, the conventional ANN described above does not provide this function. However, this function can be simulated in a neural network by using the backpropagation algorithm. The idea is to add another process so that the outputs in the final layer are used as feedback to the previous layer to update the weights, and repeat this feedback until the input layer.

Therefore, a *backpropagation neural network* or a BP-ANN consists of two major processes: (1) a conventional feedforward process which computes outputs at each layer starting from the input layer; (2) a backpropagation process where the weights are updated at each layer starting from the output layer, an attempt to improve the classification accuracy.

In order to formulate the backpropagation algorithm, the following notations are defined:

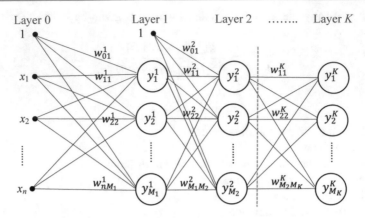

Fig. 9.16 A K-layer backpropagation neural network

- k denotes the kth layer of a network: $k = 0, 1, 2, \ldots, K$. Layer 0 is the input and layer K is the output.
- M_k denotes the number of nodes at layer k, $k = 1, 2, \ldots, K$.
- $\mathbf{x} = (x_1, x_2, \ldots, x_n)$ stands for a training sample data.
- w_{ij}^k stands for the weight of connection between node i at layer $k - 1$ and node j at layer k, $k = 1, 2, \ldots, K$.
- $net_j^k = \sum_{i=0}^{M_{k-1}} w_{ij}^k y_i^{k-1}$ stands for the output or the weighted sum of jth node of layer k, $j = 0, 1, 2, \ldots, M_k$, $k = 1, 2, \ldots, K$.
- $y_j^k = R(net_j^k)$ stands for the activated or thresholded output from the jth node of layer k, $j = 0, 1, 2, \ldots, M_k$, $k = 1, 2, \ldots, K$.

The notations are shown in the following backpropagation neural network.

The BP-ANN uses the same MSE and steepest gradient descent optimization as in the conventional ANN described earlier from (9.9) to (9.11). Assume t_j is the true output of node j at the final layer, and then the total squared error of the BP network in Fig. 9.16 is given by (9.17).

$$E = \frac{1}{2} \sum_{j=1}^{M_K} \left(y_j^K - t_j \right)^2 \tag{9.17}$$

The backpropagation algorithm starts from estimating and updating the weights of the final layer K, and then propagates the same estimating and updating procedure back to layer $K - 1$, $K - 2$, $\ldots$, until layer 1.

9.6.2 Layer K Weight Estimation and Updating

- The weights of layer K are given by w_{ij}^K, $i = 1, 2, \ldots, M_{K-1}$; $j = 1, 2, \ldots, M_K$.
- To estimate w_{ij}^K, we compute the partial derivatives $\frac{\partial E}{\partial w_{ij}^K}$.
- We will make use of the fourth property of $R(s)$ in (9.15) for the following computations.
- Remember every y_j^k is a $R(s)$ function: $y_j^k = R(net_j^k)$.
- Figure 9.17 shows the connection with weight w_{ij}^K (red line) in a BP network.

Therefore, the computation of $\frac{\partial E}{\partial w_{ij}^K}$ is simple, because among the M_K terms in (9.17), only the jth term $\left(y_j^K - t_j\right)^2$ is related to w_{ij}^K, while all the other terms are irrelevant because they are not connected to node i in layer $K - 1$. Therefore, we have the following:

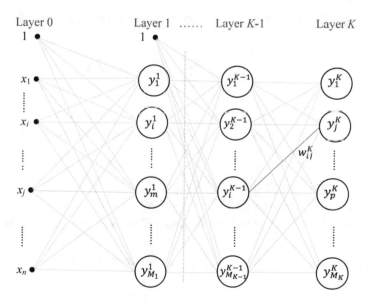

Fig. 9.17 Illustration of a connection between node i in layer $K - 1$ and node j in layer K in a K-layer BP neural network

$$\frac{\partial E}{\partial w_{ij}^K} = \frac{\partial E}{\partial y_j^K} \frac{\partial y_j^K}{\partial w_{ij}^K}$$

$$= \left(y_j^K - t_j\right) \frac{\partial y_j^K}{\partial w_{ij}^K}$$

$$= \left(y_j^K - t_j\right) \frac{\partial y_j^K}{\partial net_j^K} \frac{\partial net_j^K}{\partial w_{ij}^K} \qquad (9.18)$$

$$= \left(y_j^K - t_j\right) y_j^K \left(1 - y_j^K\right) y_i^{K-1}$$

$$= \delta_j^K y_i^{K-1}$$

where

$$\delta_j^K = y_j^K \left(1 - y_j^K\right) \left(y_j^K - t_j\right) \qquad (9.19)$$

Therefore, the weights of layer K are updated according to (9.19) as given below:

$$w_{ij}^K \leftarrow w_{ij}^K - c \frac{\partial E}{\partial w_{ij}^K} = w_{ij}^K - c\delta_j^K y_i^{K-1} \qquad (9.20)$$

where c is a positive constant. δ_j^K is equivalent to a *network sensor*, it measures if the signal of a connection should be raised or suppressed. This can be explained by (9.19) and (9.20):

- w_{ij}^K is increased (raised) if $\delta_j^K < 0$ (when $y_j^K < t_j$);
- w_{ij}^K is decreased (suppressed) if $\delta_j^K > 0$ (when $y_j^K > t_j$);
- w_{ij}^K changes little if δ_j^K is close to 0 (when y_j^K is close to t_j, 0 or 1), indicating converging;
- w_{ij}^K changes most significantly if y_j^K is very different from t_j.

9.6.3   Layer $K - 1$ Weight Estimation and Updating

- The weights of layer $K - 1$ are given by w_{mi}^{K-1}, $m = 1, 2, \ldots, M_{K-2}$; $i = 1, 2, \ldots, M_{K-1}$.
- To estimate w_{mi}^{K-1}, we compute the partial derivatives $\frac{\partial E}{\partial w_{mi}^{K-1}}$.
- The computation of partial derivative $\frac{\partial E}{\partial w_{mi}^{K-1}}$ at layer $K - 1$ is more complicated than layer K.

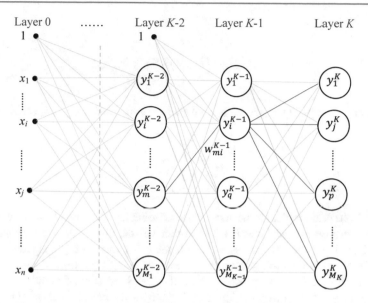

Fig. 9.18 Illustration of all connections relevant to w_{mi}^{K-1} in a K-layer BP neural network. Relevant connections are shown in red lines

- This is because w_{mi}^{K-1} connects to node i (at layer $K-1$), which is then connected to all nodes at layer K.
- The connections which are relevant to w_{mi}^{K-1} are shown in Fig. 9.18 (red lines).
- Therefore, each of the M_K terms in (9.17) is now related to the computation of $\frac{\partial E}{\partial w_{mi}^{K-1}}$.

Therefore, by using $y_j^k = R(net_j^k)$ and $dR(s)/ds = R(s)[1 - R(s)]$, the computation of $\frac{\partial E}{\partial w_{mi}^{K-1}}$ is given below by the sum of chaining derivatives:

$$
\frac{\partial E}{\partial w_{mi}^{K-1}} = \frac{1}{2} \sum_{j=1}^{M_K} \left(\frac{\partial \left(y_j^K - t_j \right)^2}{y_j^K} \frac{\partial y_j^K}{\partial net_j^K} \frac{\partial net_j^K}{\partial y_i^{K-1}} \frac{\partial y_i^{K-1}}{\partial net_i^{K-1}} \frac{\partial net_i^{K-1}}{\partial w_{mi}^{K-1}} \right)
$$

$$
= \left[\sum_{j=1}^{M_K} \left(y_j^K - t_j \right) y_j^K \left(1 - y_j^K \right) w_{ij}^K \right] y_i^{K-1} \left(1 - y_i^{K-1} \right) y_i^{K-2} \qquad (9.21)
$$

$$
\overset{(9.19)}{\Rightarrow} \left[y_i^{K-1} \left(1 - y_i^{K-1} \right) \sum_{j=1}^{M_K} \delta_j^K w_{ij}^K \right] y_i^{K-2}
$$

$$
= \delta_i^{K-1} y_i^{K-2}
$$

where

$$\delta_i^{K-1} = y_i^{K-1}\left(1 - y_i^{K-1}\right) \sum_{j=1}^{M_K} \delta_j^K w_{ij}^K \tag{9.22}$$

By repeating (9.21), it can be shown that for *any hidden layer k*, δ_i^k is given by (9.23).

$$\delta_i^k = y_i^k\left(1 - y_i^k\right) \sum_{j=1}^{M_{k+1}} \delta_j^{k+1} w_{ij}^{k+1} \tag{9.23}$$

Equation (9.23) indicates that a network sensor at layer k depends on the combined network sensors at next layer $k + 1$. In other words, during the back propagation, the weighted sum of network sensors at layer $k + 1$ has been propagated (through $dR(s)/ds$) to each network sensor at previous layer k. This kind of propagation is similar to the feedforward process where the weighted sum of network values (or signals) at layer k is propagated (through $R(s)$) to each connection at next layer $k + 1$.

The difference between the two rounds of propagation is that different propagation functions are used. In the feedforward process, the propagation function is just the activation function $R(s)$ itself, while in the backpropagation process, the propagation function is the gradient of the activation function: $dR(s)/ds$.

9.6.4 The BP Algorithm

Now that we have computed the gradients or partial derivatives of the error function E, the BP algorithm is designed as following [1]:

1. Initialize all the weights w_{ij}^k (for all i, j, k) on the network and the constant c with some small random values.
2. Input a new training data: $\mathbf{x} = (x_1, x_2, \ldots, x_n)$ from a set of N training data.
3. *Feedforward step*. Compute the outputs at each layer starting from the input layer:

$$y_j^k = R\left(\sum_{i=0}^{M_{k-1}} w_{ij}^k y_i^{k-1}\right) \quad j = 1, 2, \ldots, M_k; \quad k = 1, 2, \ldots, K.$$

4. *Backpropagation step*. Compute the network sensors at each layer starting from the output layer:

$$\delta_j^K = y_j^K\left(1 - y_j^K\right)\left(y_j^K - t_j\right) \quad \text{(for layer } K\text{)}$$

$$\text{and}\quad \delta_i^k = y_i^k\left(1 - y_i^k\right)\sum_{j=1}^{M_{k+1}} \delta_j^{k+1}w_{ij}^{k+1}\quad (\text{for } k = K-1, K-2, \ldots, 2, 1).$$

5. Update weights on the network by $w_{ij}^k \leftarrow w_{ij}^k - c\delta_j^k y_i^{k-1}$ for all i, j, k.
6. Repeat steps 2 and 5 until all the weights w_{ij}^k stop to change or stop to change significantly.

Because the BP algorithm is a steepest gradient descent algorithm, choosing the initial values for w_{ij}^k and the constant c is crucial to the performance of a network. If the initial values of w_{ij}^k are too far from the global minima, the algorithm may converge to a local minimum. Consequently, the result of class boundaries is not accurate. If the initial value of the constant c is either too small or too big, the converging progress can be very slow. In practice, several rounds of guessing the initial values of both the weights and c may be required to achieve a desirable performance.

9.7   Convolutional Neural Network

An ordinary ANN does not take raw data as input; instead, it takes features as input and classifies the data into classes based on their features. The features are computed through a separated feature extraction process (handcrafted) and are given as an n-dimensional feature vector. The reason behind this separation of feature extraction and classification is that the data dimension is usually very larger, typically from tens of thousands to millions. Direct connection of raw data to an ANN would make the network too complex and too expensive to compute with traditional computing power. Besides, the various data dimension is also a design issue for such a combined ANN.

Nowadays, with the rapid increase of computation power, it is possible to combine both the feature extraction and classification processes into a single neural network. The idea is to integrate a feature extraction network in front of an ordinary ANN. Because the feature extraction is typically done through the convolution of local filters upon an image, an ANN with feature extraction functionality is called a *convolutional neural network* or CNN for short.

9.7.1   CNN Architecture

The architecture of a CNN can be best demonstrated using the LeNet [3] in Fig. 9.19. Basically, a CNN consists of a *convolution network* in front and a *fully connected MLP* (multilayer perceptron, an ordinary ANN) at the backend.

Because each hidden unit in the convolutional network is only connected to a local neighborhood (e.g., a clock) in the input image instead of every pixel, it is also called a *locally connected* network. In contrast, in an ordinary ANN, each element

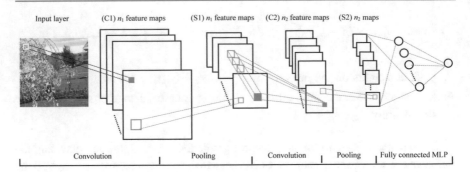

Fig. 9.19 Architecture of a CNN

of an input data is connected to each hidden unit in the network, so it is called *fully connected* network.

The convolutional network is a repeat process of *convolution* and *pooling* as shown in Fig. 9.19 [3]. Depending on the dimension of the input data, the repetitions can occur for a number of rounds. In the following, we describe the CNN architecture in detail.

9.7.2 Input Layer

- The input data are a set of training images $x_1, x_2, \ldots, x_n$.
- Each image x is a C-dimensional volume $M \times M \times C$, where M is the height and width of the image and C is the number of channels.
- For the convenience of formulation, the height and width of the images are assumed to be the same.
- Typically, the input is a RGB color image $x = x\,[i, j, k]$ and $C = 3$.
- For a gray-level image, $C = 1$.

9.7.3 Convolution Layer 1 (C1)

In a CNN, the convolution is a high-dimensional volume convolution.

- Specifically, the convolution is done by shifting a high-dimensional *volume filter* $W: N \times N \times S$ across the image as shown in Fig. 9.20a, where N is the height and width of the windowed filter and S is the number of channels of the filter.
- S can be either the same as the number of image channels C or different.
- It can be shown that a high-dimensional volume filter consists of S number of 2D filters w with size of $N \times N$.
- In practice, the convolution is done by convoluting each channel of the high-dimensional volume filter with each channel of the input image.
- Each of these 2D filters w is meant to capture a particular type of edges, shapes, or textures from the input image.

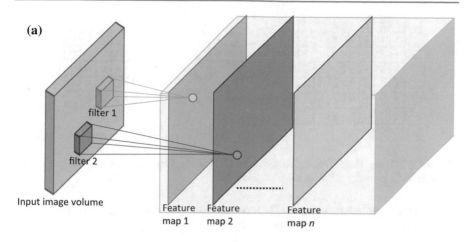

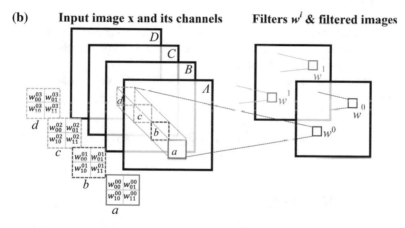

Fig. 9.20 Volume convolution. **a** Demonstration of volume filter and high-dimensional convolution; **b** demonstration of high-dimensional convolution $\mathbf{x} * w^0$ which can be done using a series of 2D convolutions. Each of the filter channels a, b, c, and d is convoluted with each of the corresponding image channels A, B, C, and D

- Figure 9.20b demonstrates how a volume convolution is done by a series of 2D convolution.
- In Fig. 9.20b, the input data is an image $\mathbf{x}$ with four channels A, B, C, and D.
- There are two high-dimensional volume filters w^0 and w^1 on the right-hand side, each of the volume filters consists of four channels a, b, c, and d, which are shown at the bottom left of Fig. 9.20 [3].
- The convolution between $\mathbf{x}$ and w^0 ($\mathbf{x} * w^0$) is done by convoluting each of the filter channels a, b, c, and d across each of the corresponding image channels A, B, C, and D.
- The convolution of image $\mathbf{x}$ with filter w^1: $\mathbf{x} * w^1$ is done the same way.

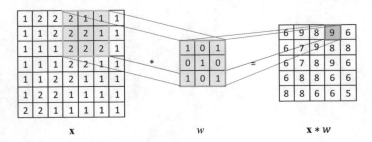

Fig. 9.21 2D convolution. An image **x** is convoluted with a filter window w and the result of the convolution is given by **x** * w at the right-hand side

- Therefore, understanding 2D convolution is the key to understand high-dimensional volume convolution.

9.7.3.1   2D Convolution

- A 2D convolution is done by sliding an $N \times N$ window w across the image **x** row by row and column by column, assuming the window slides one pixel per time and there is no padding for the moment.
- The 2D convolution of **x** * w is given by (9.24) and an example of a 2D convolution is shown in Fig. 9.21 [4].

$$X_{mn} = (\mathbf{x} * w)_{mn} = \sum_{j=0}^{N-1} \sum_{i=0}^{N-1} w[i,j] \cdot \mathbf{x}[m-i, n-j] \qquad (9.24)$$

9.7.3.2   Stride and Padding

- The dimensions of the convoluted image depend on two parameters: *stride* and *padding*.
- The *stride* determines the number of pixels the filter window shifts per time and the *padding* determines if and what the input image should be padded when the filter window is at the image boundary, e.g., 0 padding.
- If the stride value is 1 and the padding is yes, the dimensions of the convoluted image are the same as the input image.
- In Fig. 9.21, the stride is 1 and there is no padding, and therefore the convoluted image loses two pixels at both ends of each row and column.

9.7.3.3   Bias

- In a CNN, the values in a filter w are regarded as the weights for the connections between the filter and the network.

- These weights are to be learned during the training of the network.
- Therefore, a bias b is added to compensate for the estimation error.

$$X'_{mn} = (\mathbf{x} * w)_{mn} + b = b + \sum_{j=0}^{N-1}\sum_{i=0}^{N-1} w[i,j] \cdot \mathbf{x}[m-i, n-j] \tag{9.25}$$

9.7.3.4 Volume Convolution in Layer C1

- Each of the S channels of the *volume filter* is first convoluted with each corresponding channel of the input image x.
- The S filtered channels are then combined to create a 2D feature map or image f_{mn}.

$$
\begin{aligned}
f_{mn} &= \sum_{k=1}^{S} (\mathbf{x} * w[:, :, k])_{mn} + b \\
&= b + \sum_{k=1}^{S}\sum_{j=0}^{N-1}\sum_{i=0}^{N-1} w[i,j,k] \cdot \mathbf{x}[m-i, n-j, k]
\end{aligned}
\tag{9.26}
$$

9.7.3.5 Depth of the Feature Map Volume

- Multiple *volume filters* are used in a convolution layer to create a volume of feature maps.
- Each of the volume filters captures a particular type of image features.
- The number of volume filters R is called the *depth* of the feature map volume.
- The rth feature map is given by (9.27), $r = 0, 1, \ldots, R - 1$.

$$f^r_{mn} = b + \sum_{k=1}^{S}\sum_{j=0}^{N-1}\sum_{i=0}^{N-1} w^r[i,j,k] \cdot \mathbf{x}[m-i, n-j, k] \tag{9.27}$$

- Figure 9.22 shows how two volume filters w^0 and w^1 are used in layer C1 to create the two feature maps (light red) at the rightmost hand side [5].
- In the figure, the input image x is a color image with R, G, and B channels, each of the two volume filters also has three channels. The figure demonstrates the convolution of the three channels (*green*) of the first filter with a block (*yellow*) in image x. The convolutional output of the yellow block of image x is shown as the *pink* pixel in the first output image on the rightmost hand side of the figure.
- Although the input x is a 7 × 7 image, due to the stride value of the convolution is 2, the filtered output is just a 3 × 3 image. Therefore, in order to output a filtered image with the same size as the input image, we not only need to set the *stride* value as 1 but also need to *pad* the image with half the filter size.

Input volume (7x7x3)   Filter w^0 (3x3x3)   Filter w^1 (3x3x3)   Output volume (3x3x2)

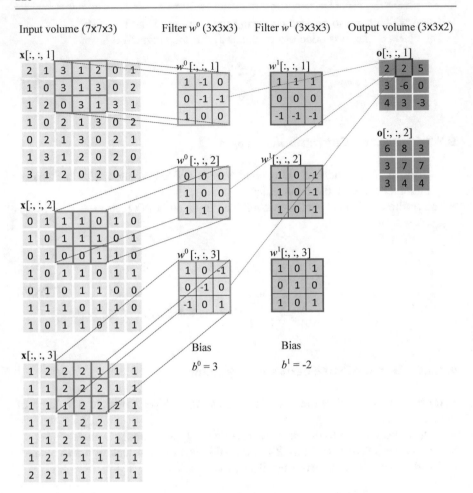

Fig. 9.22 An input image is convoluted with two volume filters w^0 and w^1. The two result feature maps are at the rightmost hand side (light red). The *stride* value is 2 and there is no padding

9.7.3.6 ReLU Activation

- The output from a volume filter or the feature map is essentially the weighted sum of the input layer.
- As in an ordinary neural network, it needs to pass an activation function.
- It has been found that in a convolution layer, the *max*(0, x) function is more effective than a sigmoid function for the activation.
- *max*(0, x) is basically a rectified linear function because it simply rectifies or refracts the negative half of $y = x$ to 0.
- Therefore, it is often called a *rectifier* and a node activated by the rectifier is also called a *rectified linear unit* or ReLU.

- So, we have

$$ReLU(x) = max(0, x) \tag{9.28}$$

- The output from node r of layer C1 is finally given as (9.29)

$$y^r_{mn} = ReLU\left(f^r_{mn}\right) \tag{9.29}$$

- The ReLU activation is usually implemented as a separate layer after the convolution layer.

9.7.3.7 Batch Normalization

In a CNN, learning rate or convergence speed is a major issue. Due to the convolution, the range of output values of the filters in each layer varies widely. In other words, the convolution has changed the original distribution of the input data, breaking the *independent* and *identically distributed* or i.i.d. assumption on input data. Worse still, each layer has to adapt to distribution drift from lower layers in order to revise its own weights. This makes the learning very inefficient and convergence very slow, especially for layers with *sigmoid* or *tanh* activation. This phenomenon is called the *internal covariance shift* or the change of data distribution from the input data distribution. In order to overcome this undesirable effect, a batch normalization procedure is introduced before the activation layer in an attempt to keep the mean and variance of the input data fixed, so that layers learn themselves more or less independent with each other. The basic idea is to normalize the input data of all layers in the network to have 0 mean and unit variance. In practice, the normalization is done to the input data or data to be activated dimension by dimension and batch by batch. Let us take a particular activation x (a single dimension of the input data), for example, there are m values from a mini-batch $B = \{x_1, x_2, ..., x_m\}$. The algorithm of the batch normalization of x is given as following [6]:

Input: Values of x from a mini-batch $B = \{x_1, x_2, ..., x_m\}$
Parameters to be learned: γ, β
Output: Batch normalized values $\{y_i = BN_{\gamma,\beta}(x_i)\}$

$$\mu_B \leftarrow \frac{1}{m}\sum_{i=1}^{m} x_i \qquad //mini-\text{batch mean}$$

$$\sigma_B^2 \leftarrow \frac{1}{m}\sum_{i=1}^{m} (x_i - \mu_B)^2 \qquad //mini-\text{batch variance}$$

$$\hat{x}_i \leftarrow \frac{x_i - \mu_B}{\sqrt{\sigma_B^2 + \varepsilon}} \qquad //sample\,normalisation$$

$$BN_{\gamma,\beta}(x_i) = y_i \leftarrow \gamma\hat{x}_i \qquad //batch\,normalisation$$

For convolution layers, the normalization is done by jointly normalizing all the activations in a mini-batch over all locations. Specifically, the pair of parameters γ and β is learnt per feature map instead of per activation [6].

By batch normalization, the values of input features to each layer are normalized into the same range, this reduces the oscillations of gradient descent when it approaches the minimum point and consequently makes it to converge faster. Another benefit of batch normalization is that it adds minor noise to each layer due to each training sample is mixed with other samples in a mini-batch, and this reduces the effect of overfitting. In practice, lower dropout rate is needed for a network with batch normalization.

9.7.4   Pooling or Subsampling Layer 1 (S1)

The feature maps' output from layer C1 usually has the same dimension as the input image. Their dimensions are too high to be connected to an ANN. Besides, the feature maps represent the finest details of the input image, these features are not as reliable. Therefore, it is tempting to downsample the feature maps so that features at a coarser level can be extracted. This is done by passing each feature map through a 2×2 subsampling function. Several types of subsampling functions can be used, such as *max()*, *average()*, L_2 *norm*, or *spectral transform* such as DWT and DCT.

For example, if the *max* function is used, the subsampling is called a *max-pooling*. Figure 9.23 demonstrates a *max-pooling* which reduces a 4×4 feature map to a 2×2 feature map.

9.7.5   Convolution Layer 2 (C2)

The outputs from the pooling layer 1 (S1) are subsampled feature maps from layer C1. New features can be computed from those feature maps by doing another round of convolution using new volume filters. The convolution procedure is the same as

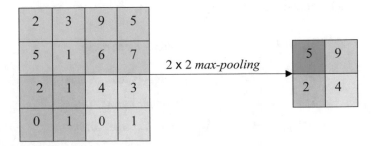

Fig. 9.23 Illustration of a max-pooling. The maximum value of each quarter block of the left image is computed as the output value of the max-pooling image at the right-hand side

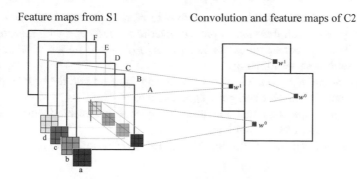

Fig. 9.24 Volume convolution in layer C2. Volume filter w^0 is convoluted with feature maps A–D while volume filter w^1 is convoluted with feature maps C–F

that in layer C1 except each volume filter in layer C2 uses different combinations of feature maps from layer S1.

For example, in Fig. 9.24, the two volume filters w^0 and w^1 are convoluted with different channels or feature maps output from layer S1. While filter w^0 is convoluted with feature maps A–D, filter w^1 is convoluted with feature maps C–F.

If the depth of S1 is R, the total number of combinations of R channels is given as follows:

$$\sum_{k=1}^{R}\binom{R}{k} = 2^R - 1 \tag{9.30}$$

where k is the number of channels in a filter in layer C2.

The convolution and pooling can be repeated for a number of rounds depending on the size of the input data. The feature maps from the final pooling layer are flattened to create a 1D feature vector, this feature vector is fed into the fully connected ANN at the backend of a CNN.

9.7.6   Hyperparameters

The performance of a CNN depends on the selection of the following hyperparameters. These parameters may be data dependent and need to be determined empirically.

- *Filter size*. The window size of the volume filter.
- *Stride*. The number of pixels per shift by the filter window at each layer.
- *Padding*. Whether padding will be used at the boundary of an input image and a feature map. What type of padding will be used, e.g., zero padding or wrap around padding.
- *Depths*. The number of channels or filters at each convolution layer.

- *Dropout-rates.* The percentage of neurons to be dropped out from each of the hidden layers at each iteration to prevent overfitting and cope with missing data.
- *Epochs.* The number of times the training algorithm will iterate over the entire training set before terminating.
- *Pooling size and function.* The size of a pooling function and the type of pooling function such as *max()*, *average()*, L_2 *norm*, etc.
- *Activation function.* The function used to generate a threshold output at each layer, such as *ReLU()*, *sigmoid*, *tanh()*, etc.
- The number of neurons in the fully connected layer of the ANN.

9.8 Implementation of CNN

To demonstrate a CNN in action, we show a high-performance CNN implementation by Oxford's Visual Geometry Group or VGGNet [7]. It has won the runner up of the 2014 ImageNet Large Scale Visual Recognition Challenge (ILSVRC) [Image-Net.org]. ImageNet is the largest hand-annotated visual dataset, and it holds image recognition competitions every year among researchers around the world. Compared with other high-performance CNN models, VGGNet is known for its simplicity because it is a *series network*.

9.8.1 CNN Architecture

The architecture of VGGNet is shown in the following list which is obtained by using Matlab code: net = vgg16; net.Layers. The 16 core layers are highlighted using bold font and are organized into five blocks: conv1—conv5. Therefore, it is often referred to as VGG16. It is a typical stacked convolution + pooling layers followed by fully connected ANN. The purpose of the softmax layer is to convert any vector of real numbers into a vector of probabilities, which correspond to the likelihoods that an input image is a member of a particular class.

01	'input'	Image Input	224 × 224 × 3 images with 'zerocenter' normalization
02	**'conv1_1'**	Convolution	64 3 × 3 × 3 convolutions with stride [1 1] and padding [1 1 1 1]
03	'relu1_1'	ReLU	ReLU
04	**'conv1_2'**	Convolution	64 3 × 3 × 64 convolutions with stride [1 1] and padding [1 1 1 1]
05	'relu1_2'	ReLU	ReLU
06	'pool1'	Max Pooling	2 × 2 max pooling with stride [2 2] and padding [0 0 0 0]
07	**'conv2_1'**	Convolution	128 3 × 3 × 64 convolutions with stride [1 1] and padding [1 1 1 1]

08	'relu2_1'	ReLU	ReLU
09	**'conv2_2'**	Convolution	128 3 × 3 × 128 convolutions with stride [1 1] and padding [1 1 1 1]
10	'relu2_2'	ReLU	ReLU
11	'pool2'	Max Pooling	2 × 2 max pooling with stride [2 2] and padding [0 0 0 0]
12	**'conv3_1'**	Convolution	256 3 × 3 × 128 convolutions with stride [1 1] and padding [1 1 1 1]
13	'relu3_1'	ReLU	ReLU
14	**'conv3_2'**	Convolution	256 3 × 3 × 256 convolutions with stride [1 1] and padding [1 1 1 1]
15	'relu3_2'	ReLU	ReLU
16	**'conv3_3'**	Convolution	256 3 × 3 × 256 convolutions with stride [1 1] and padding [1 1 1 1]
17	'relu3_3'	ReLU	ReLU
18	'pool3'	Max Pooling	2 × 2 max pooling with stride [2 2] and padding [0 0 0 0]
19	**'conv4_1'**	Convolution	512 3 × 3 × 256 convolutions with stride [1 1] and padding [1 1 1 1]
20	'relu4_1'	ReLU	ReLU
21	**'conv4_2'**	Convolution	512 3 × 3 × 512 convolutions with stride [1 1] and padding [1 1 1 1]
22	'relu4_2'	ReLU	ReLU
23	**'conv4_3'**	Convolution	512 3 × 3 × 512 convolutions with stride [1 1] and padding [1 1 1 1]
24	'relu4_3'	ReLU	ReLU
25	'pool4'	Max Pooling	2 × 2 max pooling with stride [2 2] and padding [0 0 0 0]
26	**'conv5_1'**	Convolution	512 3 × 3 × 512 convolutions with stride [1 1] and padding [1 1 1 1]
27	'relu5_1'	ReLU	ReLU
28	**'conv5_2'**	Convolution	512 3 × 3 × 512 convolutions with stride [1 1] and padding [1 1 1 1]
29	'relu5_2'	ReLU	ReLU
30	**'conv5_3'**	Convolution	512 3 × 3 × 512 convolutions with stride [1 1] and padding [1 1 1 1]
31	'relu5_3'	ReLU	ReLU
32	'pool5'	Max Pooling	2 × 2 max pooling with stride [2 2] and padding [0 0 0 0]
33	**'fc6'**	Fully Connected	4096 fully connected layer
34	'relu6'	ReLU	ReLU
35	'drop6'	Dropout	50% dropout
36	**'fc7'**	Fully Connected	4096 fully connected layer
37	'relu7'	ReLU	ReLU

38	'drop7'	Dropout	50% dropout
39	**'fc8'**	Fully Connected	1000 fully connected layer
40	'prob'	Softmax	softmax
41	'output'	Classification	crossentropyex with 'tench' and 999 other classes

VGGNet shows that the depth of the network is a critical component for good performance. The number of filters (depth) increases from 64 to 512 as it goes deeper into the network. The consecutive use of 3×3 convolutions has the effect of causing more nonlinearity as more than one ReLU functions have been applied at each stage of convolutions.

Another advantage of using consecutive convolutions is the increasing size of the receptive field. This is because two consecutive 3×3 convolutions have the effective receptive field of a single 5×5 convolution, while three-stacked 3×3 convolutions have the receptive field of a single 7×7 one [7].

9.8.2 Filters of the Convolution Layers

To validate a CNN, it is valuable to inspect and examine the internal structure of the network. Figure 9.25 shows the first 64 pretrained filters from 6 layers of VGG16 net. It can be observed that the filters typically capture the blobs, edges, regularity, directionality, and other features of an image. Filters in early layers (layers 2 and 7) typically focus on pixels, blobs, and edges, as the network goes deeper, the filters become coarser, where low-level features are organized into shapes and parts of objects.

9.8.3 Filters of the Fully Connected Layers

If the convolution layers try to capture the texture and shape features from images, the fully connected layers attempt to organize the features into objects. Figure 9.26a and b shows the first 10 channels of layers "fc6" (layer 33) and "fc7" (layer 36), respectively. This phenomenon of learning objects is more obvious in the final fully connected layer where the class names are known (using Matlab code: net.Layers (end).Classes). Figure 9.26c shows 20 channels from "fc8" (layer 39) with class names as following: goldfish, tiger shark, hammerhead shark, ostrich, great gray owl, African crocodile, mud turtle, academic gown computer keyboard, cowboy boot, accordion, cowboy hat, crane (machine), crash helmet, ambulance, analog clock, balloon, dining table, dumbbell, and acoustic guitar.

It can be observed from Fig. 9.26c that the signature patterns and shapes of these objects are well captured. More interestingly, the filters have learnt multiple copies of the same object to adapt to changes.

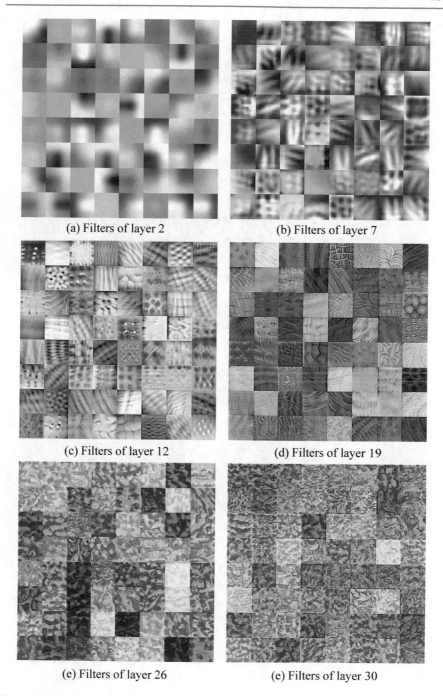

(a) Filters of layer 2

(b) Filters of layer 7

(c) Filters of layer 12

(d) Filters of layer 19

(e) Filters of layer 26

(e) Filters of layer 30

Fig. 9.25 Pretrained filters of 6 of the 13 VGG16 convolution layers

Fig. 9.26 Filters of the fully
connected layers

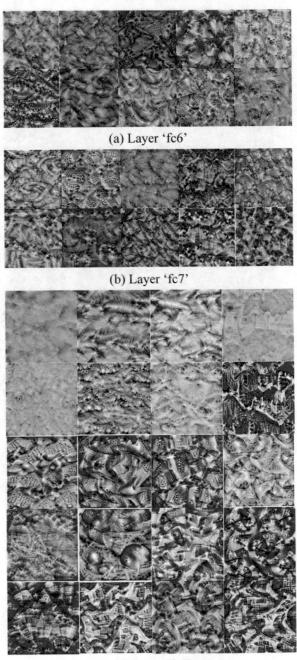

(a) Layer 'fc6'

(b) Layer 'fc7'

(c) Layer 'fc8'

9.8.4 Feature Maps of Convolution Layers

To further understand the implementation of a CNN, it is more revealing to examine the convolution process by using a real input image. A CNN is basically a combination of two components: *convolution layers* and *fully connected layers*. The convolution layers are responsible for feature extraction and the fully connected layers are responsible for the classification. The convolution component is the main powerhouse of a CNN model. Given an input image, the different filters in the convolution layers detect features such as edges, blobs, and regions, which represent eyes, ears, legs, feather, leaves, water, sand, windows, wheels, etc. The CNN does not know if they are eyes, ears, legs, etc., it learns to detect them as features by memorizing a lot of them in the input images. The fully connected layers learn how to use these features to classify the images into different classes.

One important thing to note is that due to the nature of consecutive convolution and pooling, the features leant from the CNN is evolutional or hierarchical. In other words, the CNN is a learning process from fine features to coarse features. The convolution layers learn such fine to coarse features by building on top of each other. The first layers detect edges, the next layers combine them to detect shapes, and the following layers merge shape information to infer objects such as eyes, ears, legs, etc. Figure 9.27 demonstrates this evolutional process by showing the first 64 feature maps from each of the five blocks of the lady image in Fig. 3.2: conv1_2, conv2_2, conv3_3, conv4_3, and conv5_1. It can be observed that the prominent features (hat, face) in the image are well captured by the filters. It is interesting to find that each filter captures different aspects of the image such as the surface and outline of the hat, the face, eyes, cloth, hand, background, etc. It can also be seen from the figure that the features from the first block of layers (conv1_1) are sparse and show the fine details/edges of the image, and as the network goes deeper, the features become coarser and coarser due to pooling, until the final block of layers where only the most prominent features (e.g., eyes, mouth) in the input image survive.

Figure 9.28 uses heat maps of the channels from each of the blocks to demonstrate how the prominent features of a face image have been tracked by the network. It can be seen that the eyes of the face are well tracked by the network and as the network goes deeper, the face pattern becomes coarser and coarser until it is completely blurred.

Although conventional feature extraction methods can also extract similar kind of coarse features for classification, a CNN model can combine many types of such kind of coarse features to form a set of more powerful features which lead to more accurate classification.

Overfitting is a common problem on image classification because usually there are too few training samples, resulting in a model with poor generalization

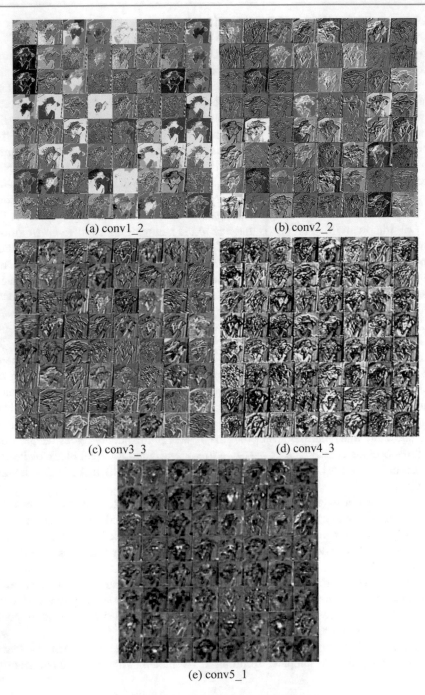

(a) conv1_2   (b) conv2_2

(c) conv3_3   (d) conv4_3

(e) conv5_1

Fig. 9.27 Feature maps from different convolution layers of VGG16

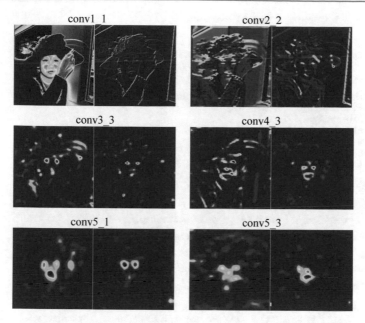

Fig. 9.28 Channels from each of the five blocks of VGG16 net

performance. One solution to overfitting is to use *data augmentation*. Data augmentation is a method to generate more training data from the current training set. It is an artificial way to boost the size of the training set, reducing overfitting.

Data augmentation is typically done by data transformations and processing, such as rotation, shifting, resizing, adding noise, contrast change, etc. It should be noted that data augmentation is only performed on the training data, not on the validation or test set.

9.8.5 Matlab Implementation

Matlab's Deep Learning Toolbox™ has a number of built-in networks which are pretrained on ImageNet, including ResNet-50, AlexNet, GoogleNet, VGG-16, and VGG-19. The following is a code scheme of using VGG-16 for image classification [8]. The code provides a step-by-step implementation of a CNN. Training images need to be first categorized and organized into subfolders, and the name of each subfolder represents the label of the image category, e.g., *bird*, *people*, *tiger*, etc.

```
% Load the Pretrained VGG-16 network
    net = vgg16();
    net.Layers;  %inspect the network architecture

% Extract and Display Feature Maps
    % Extract the filters for convolutional layer 1: conv1_1
    filter1 = net.Layers(2).Weights;
    filter1 = mat2gray(filter1);
    filter1 = imresize(filter1,5);
    figure
    montage(filter1)
    title('First convolutional layer filters')

% Prepare Training and Testing Image Sets
    % Loading images into an imageDataStore Object, e.g., rootFolder=c:\myImages,
    subFolder=cnnImages, myFolder='c:\myImages\cnnImages'
    myFolder = fullfile('rootFolder', 'subFolder');

    categories = {'cat1', 'cat2',…, 'catn'};

    imstore = imageDatastore(fullfile(myFolder, categories), 'LabelSource',
    'foldernames');

    % Split the dataset into training set (30%) and testing set (70%)
    [trainingSet, testingSet] = splitEachLabel(imstore, 0.3, 'randomize');

    % Normalise Dataset images to required size and RGB format
    imSize = net.Layers(1).InputSize;
    normTrainingSet = augmentedImageDatastore(imSize, trainingSet,
    'ColorPreprocessing', 'gray2rgb');
    normTestingSet = augmentedImageDatastore(imSize, testingSet,
    'ColorPreprocessing', 'gray2rgb');

    % Extract features from the last fully connected layer
    fcFeature = 'fc8';
    trainingFeatures = activations(net, normTrainingSet, fcFeature, ...
            'MiniBatchSize', 32, 'OutputAs', 'columns');

% Train a Multiclass SVM Classifier Using the Extracted Features
    % Get training labels from the trainingSet
    trainingLabels = trainingSet.Labels;
    classifier = fitcecoc(trainingFeatures, trainingLabels, ...
     'Learners', 'Linear', 'Coding', 'onevsall', 'ObservationsIn', 'columns');

% Test the SVM Classifier on New Images
    newImage = imread(fullfile('rootFolder', 'subFolder', 'imageName'));

    % Normalise the new images.
    normImage = augmentedImageDatastore(imSize, newImage, 'ColorPreprocessing',
    'gray2rgb');

    % Extract image features using the CNN
    imFeatures = activations(net, normImage, fcFeature, 'OutputAs', 'columns');

    % Make a prediction using the classifier
    label = predict(classifier, imFeatures, 'ObservationsIn', 'columns')
```

9.9 Summary

ANN is a powerful nonlinear classifier by layering a number of linear classifiers. However, the usage of ANN in the past has been hindered by two aspects. First, it does not appear to have the transparency as in other machine learning tools such as the Bayesian classifier, DT, and SVM. It is often regarded as a black-box-type classifier. Second, it suffers from high computation curse due to the complex optimization which involves the combination of multiple layers and a large number of nodes. However, these issues have been overcome since the introduction of CNN and more powerful computing hardware and software. Furthermore, CNN has also extended the traditional ANN from network and computing (fully connected layers) to including sensors (the convolution layers). This means that CNN is a simulation of a complete human visual system. The downside of a CNN though is the high number of parameters which needs to be determined empirically.

The introduction of CNN is a significant development to ANN and machine learning as a whole. This is because the convolution layers can be independent of the ANN and they can be connected to any other machine learning tools such as the Bayesian classifier, DT, SVM, etc. With the Matlab Deep Learning toolbox and other powerful tools, the internal structure and learning process of a CNN can now be studied at the same transparent level as other machine learning tools. Because of the transparency, the computation and learning processes can now be controlled using a step-by-step approach. Due to the development of CNN, ImageNet, and the ever increasing computation power, the future of image data mining has become much brighter.

References

1. Gose E, Johnsonbaugh R, Jost S (1996) Pattern recognition and image analysis. Prentice Hall
2. Tivive F, Bouzerdoum A (2005) Efficient training algorithms for a class of shunting inhibitory convolutional neural networks. IEEE Trans Neural Nctw 16(3):541–556
3. Theano Development Team (2019) Convolutional neural networks (LeNet). http://deeplearning.net/tutorial/lenet.html#the-full-model-lenet. Accessed in Feb 2019
4. Latysheva N (2019) Implementing your own K nearest neighbour algorithm using Python. https://cambridgespark.com/content/tutorials/convolutional-neural-networks-with-keras/index.html. Accessed in Feb 2019
5. Karpathy A (2019) Stanford CS class CS231n: convolutional neural networks for visual recognition. http://cs231n.github.io/convolutional-networks/. Accessed in Feb 2019
6. Ioffe S, Szegedy C (2015) Batch normalization: accelerating deep network training by reducing internal covariate shift. In: Proceedings of the 32nd international conference on international conference on machine learning, no 37, pp 448–456

7. Dertat A (2019) Applied deep learning—Part 4: Convolutional neural networks. https://towardsdatascience.com/applied-deep-learning-part-4-convolutional-neural-networks-584bc134c1e2. Accessed in Feb 2019
8. Mathworks (2019) Image category classification using deep learning. http://au.mathworks.com/help/vision/examples/image-category-classification-using-deep-learning.html#vision_product-DeepLearningImageClassificationExample. Accessed in Feb 2019

Image Annotation with Decision Tree 10

We may be different, but we all share a common ancestor.

10.1 Introduction

The machine learning methods we discussed so far are typically black box type of classifiers, in the sense that the decisions they make are not transparent to users. In other words, these models are neither interpretable nor comprehensible to users. Another issue with these methods is that their decision-making process is one path or non-conditional process, which means that there are no alternatives when the original decision was not appropriate.

Human beings, however, tend to make decisions in a step-by-step and hierarchical way. For example, when we look at an image with complex patterns, we tend to first organize the different patterns into groups using the most prominent attribute or feature, then go further to identify the objects we are interested, and analyze them in detail using other types of attributes or features. This kind of hierarchical and step-by-step analysis is repeated until we are satisfied.

In machine learning, this kind of intuitive, hierarchical, and step-by-step analysis can be modeled using a *decision tree* or DT. DT is a "divide-and-conquer" approach to learn classification from a set of training samples. A DT is built from a training dataset by recursively dividing the dataset into several subsets based on the possible values of a selected attribute. The procedure starts at the root node and continues until all the instances of a subset have the same class label or there is no other attribute left to divide them.

A DT is typically built upside down with its root at the top. Figure 10.1 shows an example of a DT on image classification [1, 2]. On the DT, an *internal node* (with outgoing branches) is labeled with an input feature or a selected attribute. The branches coming from a node are labeled with each of the possible values of the selected attribute. Each *leaf node* (without outgoing branch) of the tree is labeled with a class or a probability distribution over the classes.

© Springer Nature Switzerland AG 2019
D. Zhang, *Fundamentals of Image Data Mining*, Texts in Computer
Science, https://doi.org/10.1007/978-3-030-17989-2_10

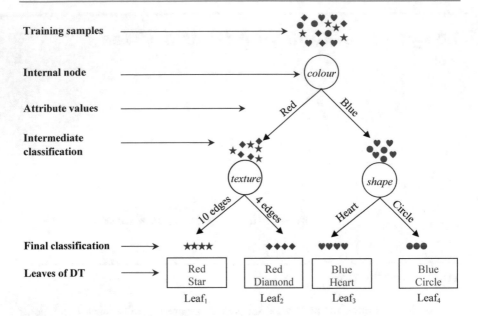

Fig. 10.1 A DT for image classification

The following is a list of terminologies associated with a DT:

1. *Root Node*. It represents the entire dataset and it is to be divided into two or more homogeneous sets.
2. *Internal Node/Decision Node*. It is a node which can be split into two or more sub-nodes.
3. *Leaf Node/Terminal Node*. It is a node which cannot be split into sub-nodes.
4. *Branch/Sub-tree*. An edge coming out of a node is called a branch and the section under a branch is called a sub-tree.
5. *Parent and Child Node*. A node which is divided into sub-nodes is called parent node and a node under a parent node is called a child node.
6. *Splitting*. It is a process of dividing a node into two or more sub-nodes.
7. *Pruning*. It is a process of removing unwanted sub-nodes and branches. It is the opposite process of splitting.

Depending on the type of attribute values, a DT can be either a *classification tree* or a *regression tree*. A classification tree takes a *discrete* set of attribute values and the predicted outcomes are the class labels to which the data belong, while a regression tree takes *continuous* attribute values and the predicted outcomes are real numbers.

Quinlan [3] first formulated a DT algorithm called ID3 (Iterative Dichotomiser 3) which only accepts discrete features. ID3 is later extended to C4.5 [4] which accepts both discrete and continuous features. In the following, we describe the characteristics of different types of DTs.

10.2   ID3

The ID3 algorithm begins with a training dataset T and an attribute set A as the root node. It then checks every unused attribute of the attribute set A and calculates the entropy $info(T)$ (or information gain $IG(T)$) of that attribute. It then selects the attribute which has the smallest entropy (or largest IG) value. The set T is then split into subsets by the selected attribute. The above procedure is repeated on each subset until there is no unused attribute or the subset is homogeneous (all instances in the subset are from the same class). ID3 only accepts data with discrete or nominal values. The algorithm of the ID3 can be summarized as follows:

ID3 (T, A) {

1. Create a root node *Root*
2. If all instances in *Root* belong to the same class C or A is empty

   2.1. Stop and return *Root* with label $= C$

3. Else

   3.1. Select an attribute A_i with possible values $A_i^1, A_i^2, \ldots, A_i^n$
   3.2. Partition T into subsets $T_1, T_2, \ldots, T_n$ according to the values of A_i
   3.3. For each T_j

       3.3.1. Add a new branch under *Root* to connect T_j with *Root*
       3.3.1. If T_j is homogeneous

           3.3.1.1.   Then, below this new branch add a leaf node with label $= A_i^j$

       3.3.1. Else, below this new branch add the sub-tree ID3 $(T_j, A - \{A_i\})$

}

This algorithm, in fact, is a *general DT algorithm*. Central to a DT algorithm is step 3.1, which requires an attribute selection criterion or a splitting criterion. This splitting criterion determines how the tree looks like and the performance of a DT as well.

10.2.1   ID3 Splitting Criterion

Because the split criterion is critical to the success of a DT, variety of criteria have been proposed. The rule is to select an attribute which reduces the maximum amount of uncertainty in data, because the higher the uncertainty in data is, the more difficult it is to predict the class of an instance. Intuitively, this is equivalent to selecting the most useful or most telling attribute to make a decision. Information gain is a statistical measurement of reducing the uncertainty in data. Therefore, in ID3, the attribute which gives the highest information gain is selected as the test attribute.

Information gain of an attribute measures how much information we can save or gain if it is selected to split the training set. Mathematically, it is measured as the *difference* between information needed to classify an instance before and after the attribute splits the training dataset.

Information before the splitting:

- Given a training set $T = \{(\mathbf{x}_1, y_1), (\mathbf{x}_2, y_2), \ldots, (\mathbf{x}_N, y_N)\}$, where $\mathbf{x}_i$ is the data or sample, y_i is the class label for $\mathbf{x}_i$, and $y_i \in \{C_1, C_2, \ldots, C_m\}$.
- Instances in T are characterized with the set of attributes $A = \{A_1, A_2, \ldots, A_n\}$.
- Each attribute A_i has possible values $A_i^1, A_i^2, \ldots, A_i^{n_i}$.
- The probability that an instance of T belongs to class C_j is given as

$$P_j = \frac{\|C_j\|}{\|T\|} \tag{10.1}$$

where $\|C_j\|$ is the number of instances in C_j.

To classify an instance in T, the information needed (or the *entropy*) is given as,

$$info(T) = -\sum_{j=1}^{j=m} P_j \times \log P_j \tag{10.2}$$

The negative sign before the sum is to make the information a positive value; this is because $P_j \leq 1$ and $\log P_j \leq 0$. Generally, *entropy* refers to *disorder* or *uncertainty* in a dataset; a smaller $info(T)$ means a more predictable class. This is consistent with our understanding of information. An event with higher chance of occurrence carries little to zero information such as sunrise/sunset, while an event with less chance of occurrence carries more information such as rain/sunshine. In English language, frequent words such as "a", "the", and "this" carry almost zero information, while rare words such as "Delphi", "nirvana", and "dialectics" carry a lot of information. In terms of a dataset, a data source (or a class) with higher probability value carries less information than a data source with lower probability value. Therefore, a DT learning algorithm attempts to split T into subsets so that the expected information needed is minimized after a split.

Information after the splitting:

Suppose, an attribute A_i has n_i nominal values such as $A_i^1, A_i^2, \ldots, A_i^{n_i}$. If attribute A_i is selected at the current node, it splits the training set T into $T_i^1, T_i^2, \ldots, T_i^{n_i}$. After the splitting, the *expected information* is calculated as

$$E(A_i) = \sum_{j=1}^{j=n_i} \frac{\|T_i^j\|}{\|T\|} \times info(T_i^j) \qquad (10.3)$$

The information gain is the difference between info(T) and E(A$_i$):

$$IG(A_i) = info(T) - E(A_i) \qquad (10.4)$$

The attribute which gives the highest *IG* is chosen for splitting the training set. Because *info(T)* in (10.2) is the same for all attributes, the attribute A_i which gives the highest gain has lowest expected information $E(A_i)$. Therefore, the attribute which leads to the least expected information is selected.

10.3   C4.5

C4.5 build a DT from a training dataset in the same way as ID3, except C4.5 has made a number of improvements to ID3:

- Use *gain ratio (GR)* instead of *IG* to build a better DT.
- Accept both *continuous* and *discrete* attributes. For continuous attributes, C4.5 creates a threshold and then splits the list into those whose attribute value is above the threshold and those that are less than or equal to it.
- Handle *incomplete data* points. C4.5 allows attribute values to be marked as "?" for missing data.
- Apply different *weights* to the attributes.
- Overcome *over-fitting* problem by a bottom-up *pruning*. C4.5 goes back through the tree once it has been created and removes branches that are deemed unnecessary by replacing them with leaf nodes.

10.3.1   C4.5 Splitting Criterion

In ID3, *IG* is used as splitting criterion. However, the disadvantage of using *IG* as splitting criterion in ID3 is that it favors the highly branching attributes, that is, the attributes which have a large number of possible values. Let us think about the extreme case where the instance ID is used as an attribute. Say, it is denoted as A_{ID}, and it has a distinct value for each instance. If A_{ID} is used to split the dataset T, each subsequent subset will have only one instance. According to (10.3), $E(A_{ID})$ is zero because each subset has zero entropy (information). Therefore, $IG(A_{ID})$ will be the highest and A_{ID} will be selected. However, such a selection tells nothing about the nature of the decision and leads to no classification at all. To reduce the effect of high branching factor on information gain, a modified measure called *gain ratio* (*GR*) or *normalized IG* is proposed by Quinlan in C4.5. The gain ratio is defined as *IG* normalized by split information [4]

$$GR(A_i) = \frac{IG(A_i)}{splitInfo(A_i)} \tag{10.5}$$

where *splitInfo*(A_i) is calculated based on the proportion of each subset resulted from the split using attribute A_i, regardless of the class information inside each subset. Specifically, it is defined as

$$splitInfo(A_i) = -\sum_{j=1}^{j=n_i} \left(\frac{\|T_i^j\|}{\|T\|} \times \log\left(\frac{\|T_i^j\|}{\|T\|} \right) \right) \tag{10.6}$$

It can be observed that the higher the number of attribute values of A_i, the larger the magnitude of *splitInfo*(A_i) and the lower its gain ratio. Therefore, using gain ratio, the high branching behavior is penalized.

10.4   CART

In machine learning, a DT can be either *classification tree* or a *regression tree*. For a classification tree, the predicted outcome is a class such as tree, tiger, water, etc., while for a regression tree, the predicted outcome is a real number, such as stock price, queueing time, etc.

CART stands for Classification And Regression Tree; it is an umbrella term used to cover both classification DT and regression DT. It was first introduced by Breiman et al. in 1984 [5]. A CART tree is a *binary* DT that is constructed by splitting a node into two child nodes repeatedly, beginning with the root node that contains the entire training data. The splits are done using the twoing criteria and the obtained tree is pruned by cost–complexity technique. CART can handle both numeric and nominal attributes, and it can also handle outliers.

10.4.1   Classification Tree Splitting Criterion

A CART uses splitting criteria called the *twoing criteria* for a classification tree, which is defined as (10.7)

$$\Delta i(t) = \frac{P_L P_R}{4} \left(\sum_{j=1}^{m} |p(j|t_L) - p(j|t_R)| \right)^2 \tag{10.7}$$

where

- t is the node to be split,
- $\Delta i(t)$ indicates the *impurity* of node t,
- P_L is the proportion of data split into the left node t_L and similar for P_R,
- $P_L = ||t_L||/||t||$ and $P_R = ||t_R||/||t||$, where t is the parent node of t_L and t_R, $||t||$ is the total number of data in node t,
- $p(j|t_L)$ is the proportion of data belonging to class j at the left node t_L, and
- m is the number of classes in the training set.

The twoing criteria measure the difference between the two split nodes, and a split is achieved by maximizing the difference or $\Delta i(t)$.

Gini impurity can also be used to define a splitting criterion for a classification tree. First, the Gini index (*GI*) or *GI*s are computed for both the left split node t_L and right split node t_R, which are given in (10.8) and (10.9), respectively.

$$GI(t_L) = 1 - \sum_{j=1}^{m} [p(j|t_L)]^2 \qquad (10.8)$$

$$GI(t_R) = 1 - \sum_{j=1}^{m} [p(j|t_R)]^2 \qquad (10.9)$$

It can be observed from (10.8) and (10.9) that

- A *GI* is maximum or $GI = (1 - 1/m)$ when records in a node are equally distributed among all classes, indicating maximum uncertainty.
- A *GI* is minimum or $GI = 0$ when all records in a node belong to one class, indicating minimum uncertainty.

A split is achieved by minimizing the *Gini impurity* which is defined as (10.10)

$$iG(t) = [||t_L|| \cdot GI(t_L) + ||t_R|| \cdot GI(t_R)]/||t|| \qquad (10.10)$$

The Gini impurity splitting algorithm works faster than twoing splitting algorithm; however, the twoing splitting criterion builds a more balanced DT and offers a superior performance on complex classification such as multi-class and noisy data.

10.4.2 Regression Tree Splitting Criterion

A regression tree is also called a prediction tree. Instead of identifying the class label for a training or unknown data, a regression tree predicts the likely target value of the data. For regression tree, the splitting criterion is typically given by the mean squared error (*MSE*). Given a training set $T = \{(\mathbf{x}_1, y_1), (\mathbf{x}_2, y_2), \ldots, (\mathbf{x}_N, y_N)\}$, where $\mathbf{x}_i$ is the data or sample and y_i is the target value for data $\mathbf{x}_i$. The splitting is determined by the two nodes which give the smallest *MSE*:

$$MSE = \frac{1}{N_L} \sum_{i=1}^{N_L} (y_i - \hat{y}_L)^2 + \frac{1}{N_R} \sum_{j=1}^{N_R} (y_j - \hat{y}_R)^2 \qquad (10.11)$$

where

- N_L and N_R are the total number of samples falling into the left split node and the right split node, respectively,
- $\hat{y}_L$ and $\hat{y}_R$ are the prediction values for the left and right split nodes, respectively,
- $\hat{y}_L$ is typically given as the result of a *regression* from the data falling into the left node: $\hat{y}_L = \hat{f}_L(\mathbf{x}_i) = \boldsymbol{\beta}^T \cdot \mathbf{x}_i + b$.

A *regression* tree is basically a *piecewise linear* approximation of a dataset in space. Figure 10.2 demonstrates a contrast between a global linear regression and a regression tree approximation on one variable dataset. In the figure, the green line shows an approximation from a global linear regression, while the red lines represent a regression tree approximation of the same data. It can be observed that the regression tree gives a much closer approximation than the global linear regression model.

10.4.3   Application of Regression Tree

The prediction value $\hat{y}_L$ in (10.11) can also be estimated by the *mean* of the left node $\hat{y}_L = \frac{1}{N_L} \sum_{i=1}^{N_L} y_i$ (similar for $\hat{y}_R$). The tree built in this way provides a *piecewise constant* approximation of the data. The mean prediction model is a much faster method to build a regression tree.

Figure 10.3 shows a mean prediction tree for predicting median house price of California based on the two variables: latitude and longitude. The actual data map and the tree partitions are shown in Fig. 10.4. It can be seen that the finer partitions are concentrated at the darker areas.

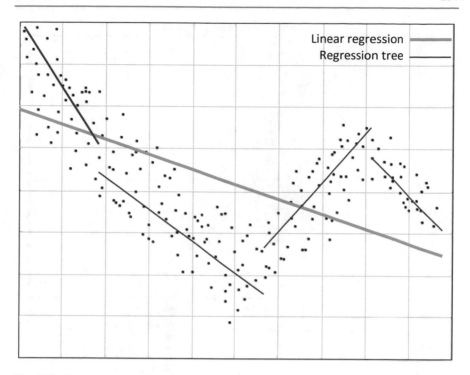

Fig. 10.2 Contrast between linear regression and regression tree

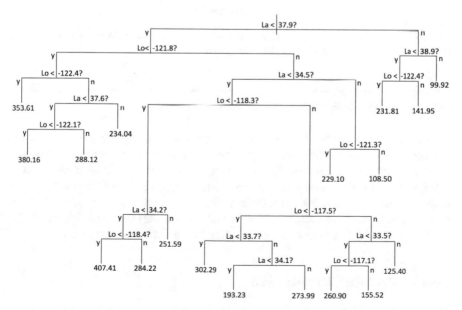

Fig. 10.3 A regression tree for predicting median house price ('000) in California from the geographic coordinates of Fig. 10.4. Legend: La = latitude, Lo = longitude, y = yes, n = no

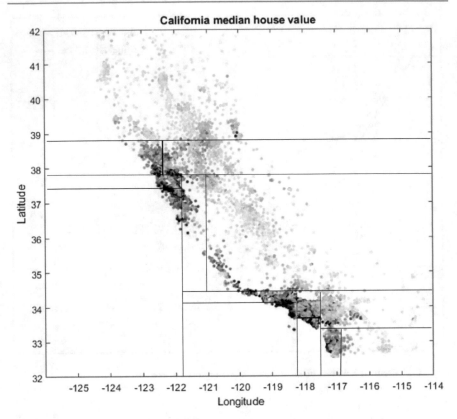

Fig. 10.4 Data map of actual median house prices in California and the tree partition of the data map, the darker the color, the higher the house value

10.5   DT for Image Classification

Images are complex data. A typical image usually has multiple regions/objects and has multiple interpretations. Therefore, the first step is to segment an image into individual regions and represent each region with an n-dimensional feature vector: $\mathbf{x} = (x_1, x_2, \ldots, x_n)$. For color images, each image region is represented with a color feature vector $\mathbf{x}_C$ and a texture feature vector $\mathbf{x}_T$. Because certain types of image regions can be best described by both color and texture, the third feature is also created by combining both color and texture into a single feature vector $\mathbf{x}_{CT}$.

10.5.1   Feature Discretization

The three types of features $\mathbf{x}_C$, $\mathbf{x}_T$, and $\mathbf{x}_{CT}$ are all continuous features; in order to classify these regions using a classification DT, the features need to be quantized into discrete values using a vector quantization (VQ) technique. The idea of a VQ technique is to cluster similar image regions into clusters which are then assigned with nominal values such as 0, 1, 2, ..., K. These nominal values correspond to class labels such as sky, water, grass, firework, tiger, etc., which are used for DT classification. Common VQ algorithm is the LBG algorithm [6] which is given as follows:

LBG (T, K, ε) {
Input: $T = \{\mathbf{x}_i \in R^n, i = 1, 2, \ldots, N\}$
Output: $C = \{c_j \in R^n, j = 1, 2, \ldots, K\}$

1. Initiate a codebook $C = \{c_j \in R^n, j = 1, 2, \ldots, K\}$
2. Set $D_0 = 0$ and $k = 0$
3. Classify the N training vectors into K clusters $T_q (q = 1, 2, \ldots, K)$ and classify $\mathbf{x}_i$ to T_q if the distance $d(\mathbf{x}_i - c_q) < d(\mathbf{x}_i - c_j)$ for all $j \neq q$
4. Update cluster centroids c_j by $c_j = \frac{1}{|T_j|} \sum_{\mathbf{x}_i \in T_j} \mathbf{x}_i (j = 1, 2, \ldots, K)$
5. Set $k \leftarrow k + 1$ and compute the distortion $D_k = \sum_{j=1}^{K} \sum_{\mathbf{x}_i \in T_j} d(\mathbf{x}_i - c_j)$
6. If $(D_{k-1} - D_k)/D_k > \varepsilon$ *(a small positive number)*, repeat steps 3–5
7. Return the codebook $C = \{c_j \in R^n, j = 1, 2, \ldots, K\}$

}

By applying the LBG VQ algorithm on $\mathbf{x}_C$, $\mathbf{x}_T$, and $\mathbf{x}_{CT}$, three codebooks or visual dictionaries $V_i (i = 1, 2, 3)$ are created:

$$V_i = \left\{ v_1^i, v_2^i, \ldots, v_{n_i}^i \right\}, \quad i = 1, 2, 3 \tag{10.12}$$

where v_j^i represents a code word of V_i and n_i is the total number of code words in V_i.

For each image region R in the training dataset, it is represented as three discrete attribute values as follows:

$$R = (ind_1, ind_2, ind_3) \tag{10.13}$$

and

$$ind_i = \arg \min_j (dist(v^i, v_j^i)) \tag{10.14}$$

where

- $dist(v^i, v_j^i)$ is the distance between feature v^i and code word v_j^i,
- v^i is one of the feature vectors ($\mathbf{x}_C$, $\mathbf{x}_T$ or $\mathbf{x}_{CT}$) of R,

- v_j^i is the jth value of attribute V_i, and
- ind_i is an index value of attribute V_i and it is an integer between 0 and n_i.

Therefore, each image region in the training set is now associated with three discrete attributes $A_i = V_i(i = 1, 2, 3)$ which are ready for building a DT.

10.5.2   Building the DT

With a training set of image regions T and a set of visual attributes A, an image classification DT can be built using the following algorithm [1]:

1. If all training regions of T belong to the same class C,

    1.1. The tree is a leaf node with the outcome C.
    1.2. Stop.

2. If the regions of T belong to more than one class but there is no attribute to separate them,

    2.1. The tree is a leaf node. The outcome is determined as follows.

        2.1.1. If there is a single majority class in T, the outcome is that class.
        2.1.2. Else, the outcome is the majority class of the parent node.

    2.2. Stop.

3. If the regions of T belong to more than one class and there are one or more attributes to separate them

    3.1. Create an internal node.
    3.2. Calculate the *IG* or *gain ratio* for each attribute.
    3.3. Select the attribute A_i with the highest gain ratio: $A_i = \{v_1^i, v_2^i, v_3^i, \ldots, v_{n_i}^i\}$
    3.4. Use A_i as the test attribute for the internal node.
    3.5. Split the training set T into subsets: $T_i^0, T_i^1, T_i^2, T_i^3, \ldots, T_i^{n_i}$, where image regions in T_i^j have attribute value v_i^j.
    3.6. Remove attribute A_i from the attribute list.
    3.7 Repeat from Step 1 for each T_i^j.

A DT generated from the above algorithm can have nodes with isolated or noisy samples. A common practice is to include a pre-pruning procedure after each splitting to remove those nodes with noisy samples. The following pre-pruning procedure can be included as step 3.7 in the above algorithm, and name the original step 3.7 as 3.8.

3.7 Pre-pruning [7]:

3.7.1 Calculate the probability P'_i for every class in subset T_i^j ($j = 0, 1, 2, \ldots, n_i$) as

$$P'_i = \frac{\|p_j\|}{\|p_i\|} \qquad (10.15)$$

where p_i and p_j are the number of instances of class C_i in T and T_i^j, respectively.

3.7.2 Remove those samples from subset T_i^j whose class probability P'_i is less than a threshold k.

3.7.3 Remove T_i^j if it is an empty subset after sample removal.

Figure 10.5 shows a DT learnt from a dataset of 570 image regions which have been quantized into 19 classes (30 images/class), and the pre-pruning threshold is $k = 0.1$ [8]. The meanings of the leave labels are as follows: A = *Grass*, B = *Forest*, C = *Sky*, D = *Sea*, E = *Flower*, F = *Sunset*, G = *Beach*, H = *Firework*, I = *Tiger*, J = *Fur*, K = *Eagle*, L = *Building*, M = *Snow*, N = *Rock*, O = *Bear*, P = *Night*, Q = *Crowd*, R = *Butterfly*, S = *Mountain*, and U = *Unknown*.

A DT generated from the above algorithm can still be very complex and imbalanced. A post-pruning procedure is usually applied after the initial tree has been generated; this can be added as step 4 in the above DT algorithm. The post-pruning is a procedure to *remove those isolated branches* and merge them with neighboring nodes.

1. Post-pruning [7]:

   a. If for more than one value of the attribute A_i, the outcome class labels are identical, i.e., C_i, then all the leaf nodes corresponding to these attribute values are merged as a single leaf node labeled with class C_i.
   b. If the outcomes for all the possible values of attribute A_i are identical, i.e., C_i, then the *sub-tree* rooted at A_i is replaced with a single leaf node with C_i as an outcome.

Figure 10.6 shows the DT from Fig. 10.5 after post-pruning; it is a much simpler and more robust DT [7, 8].

10.5.3 Image Classification and Annotation with DT

Once the DT is generated from a training dataset, a set of rules or a *DT model* can be formulated from the DT by traversing the tree from the root to each of the leaf

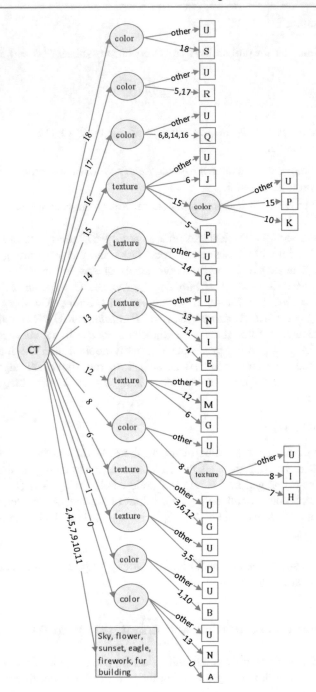

Fig. 10.5 A DT for image classification without post-pruning, CT = color and texture

nodes. This DT model is to be used as an image classifier. The following is the DT model formulated from the DT shown in Fig. 10.6.

```
IF CT is 1 (or 2, 3, 4, 5, 6, 7, 9, 10, 11, 12, 14, 15, 16, 17) THEN
{
        Region is Forest (or Sky, Sea, Flower, Sunset, Beach, Firework, Fur, Eagle, Building,
        Snow, Bear, Butterfly, Crowd, Mountain respectively)
}
ELSE IF CT is 0 AND
{
        IF color is 13 THEN Region is Rock
        ELSE Region is Grass
}
ELSE IF CT is 8 AND
{
        IF color is 8 AND
                IF texture is 7 THEN Region is Firework
                ELSE Region is Tiger
        ELSE Region is Tiger
}
ELSE IF CT is 13 AND
{
        IF texture is 4 THEN Region is Flower
        ELSE IF texture is 11 THEN Region is Tiger
        ELSE Region is Rock
}
ELSE IF CT is 15 AND
{
        IF texture is 9 THEN Region is Fur
        ELSE IF TEXTURE IS 15 AND
                IF color is 10 THEN Region is Eagle
                ELSE Region is Night
        ELSE Region is Night
}
END
```

Given a new or unknown image, it is also segmented into regions using the same algorithm as that used by the training dataset. Each region is then represented as three discrete attribute values: $R = (ind_1,\ ind_2,\ ind_3)$, using the learnt visual dictionaries. Each of the regions R is then analyzed using the DT model as shown in the above and is classified into a class.

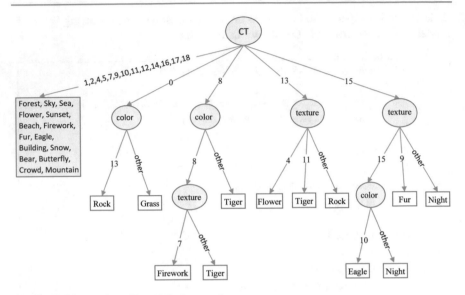

Fig. 10.6 The DT from Fig. 10.5 after pruning

As can be observed, a DT model is completely transparent and comprehensible to a human user. A classifier working in this way can be easily modified and fine-tuned to adapt to the data. The other advantage of the DT model is that issues can be found and corrected at learning stage without much difficulty.

10.6 Summary

DT is a powerful image classification tool. Due to its hierarchical nature and piecewise approximation of data, it offers a middle ground between *generative* and *discriminative* approaches. Compared with other classification tools, DT has a number of advantages.

- DT is a tool known for its simplicity, transparency, and comprehensibility.
- DT can handle both numeric and categorical attributes.
- DT can handle both noisy and missing data.
- DT offers an intuitive and step-by-step analysis based on selected attributes.
- DT does not require complex computation.
- DT generates rules which are easy to interpret.

A DT may grow too complex and imbalanced due to noise, fragmentation, and missing data. Therefore, a pruning mechanism is essential to a DT algorithm. Common practice is to apply a post-pruning technique after the tree has been generated. However, many misplaced instances would have a better chance to be classified into more appropriate nodes if a pre-pruning is applied during the tree building. A well-designed pruning scheme can improve the DT performance significantly.

According to the Ockham's Razor principle, the simplest rule that is consistent with all observations is the best. In terms of DT, it means that the smallest decision tree that correctly classifies all the training examples is the best.

References

1. Islam M (2009) SIRBOT—semantic image retrieval based on object translation. PhD thesis, Monash University
2. Zhang D, Islam M, Lu G (2012) A review on automatic image annotation techniques. Pattern Recognit 45(1):346–362
3. Quinlan J (1986) Induction of decision trees. In: Springer machine learning, pp 81–106
4. Quinlan J (1993) C4.5: programs for machine learning. Morgan Kaufmann, Los Altos, California, USA
5. Breiman L et al (1984) Classification and regression trees. Taylor & Francis
6. Linde Y, Buzo A, Gray R (1980) An algorithm for vector quantizer design. IEEE Trans Commun 28(1):84–95
7. Liu Y, Zhang D, Lu G (2008) Region-based image retrieval with high-level semantics using decision tree learning. Pattern Recognit 41(8):2554–2570
8. Liu Y (2006) Region-based image retrieval with high-level semantics. PhD thesis, Monash University

Part IV
Image Retrieval and Presentation

Quality over quantity.

Image Retrieval (IR) is a set of techniques to retrieve images from a very large image database to meet a user's expectation, e.g., find a group of "computer" images on the Web. The database is either classified or not. Even if an image database is classified, to retrieve similar images from the database is still a challenging task. This is due to two reasons. First, the database is typically very large, a commercial image database typically numbers from millions to hundreds of millions of images. Second, the classification is not perfect given current classification technology. In fact, the classification accuracy is usually quite low especially for a very large database with vast varieties of images.

IR is an intensive research area, research on IR mostly focuses on four major topics: feature extraction, image indexing, image ranking, and image presentation. The first topic has been comprehensively covered in Part II; in this part of the book, we focus on the remaining three topics.

Image indexing is to organize the image database into data structure such as a list or a tree to facilitate fast search. Image ranking is a technique to assess the similarity between database images, so that given a query image, images which are similar to the query can be identified and retrieved. Image presentation is a method of presenting similar images to the user so that browsing the retrieved images is the most convenient and efficient.

Image Indexing

11

Order is our favourite, but the truth is beyond order.

11.1 Numerical Indexing

11.1.1 List Indexing

In numerical indexing, each image I in the database has been represented as a k-dimensional feature vector: $\mathbf{x} = (x_1, x_2, \ldots, x_k)$. The simplest way of a numerical indexing is to create a list of (*image_id*, $\mathbf{x}$) as shown in the following:

$$(I_1, \mathbf{x}_1) = [I_1, (x_{11}, x_{12}, \ldots, x_{1k})]$$
$$(I_2, \mathbf{x}_2) = [I_2, (x_{21}, x_{22}, \ldots, x_{2k})]$$
$$\vdots$$
$$(I_N, \mathbf{x}_N) = [I_N, (x_{N1}, x_{N2}, \ldots, x_{Nk})]$$

where N is the total number of images in the database.

A list is simple and useful for a small image database, however, it's impossible to use it for a very large commercial image database, because it would take a long time to search the entire list of millions even billions of images. Therefore, a more efficient data structure is needed to index large image databases. One of the simplest yet efficient data structure is the *k-d tree* indexing.

11.1.2 Tree Indexing

The simplest *k-d* tree is a binary tree. A binary *k-d* tree is an algorithm of repeatedly splitting the database into two subsets by cyclically dividing the k dimensions of the data. Given N number of k-dimensional feature vectors: $\mathbf{x}_1, \mathbf{x}_2, \ldots, \mathbf{x}_N$, a *k-d* tree first divides the N data into two sets or branches of approximately equal size

© Springer Nature Switzerland AG 2019
D. Zhang, *Fundamentals of Image Data Mining*, Texts in Computer
Science, https://doi.org/10.1007/978-3-030-17989-2_11

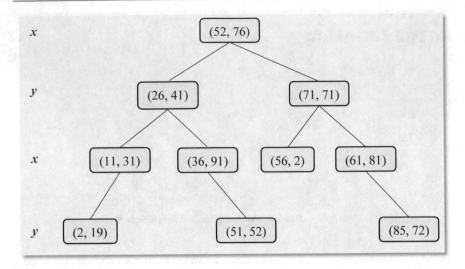

Fig. 11.1 A *k-d* tree for a 2D dataset of 10 data

according to the values of the first dimension of the N feature vectors. Next, each of the left and right branches are further divided into two subbranches of approximately equal size according to the values of the second dimension of the feature vectors. This division process continues until the kth dimension when the next division returns to the first dimension. The cycle goes on until no node is divisible any more.

Figure 11.1 shows an example of a *k-d* tree for a 2D dataset. Given a 2D dataset of 10 data: (56, 2), (36, 91), (52, 76), (2, 19), (11, 31), (61, 81), (85, 72), (71, 71), (51, 52), (26, 41), the binary *k-d* tree for this dataset is shown in Fig. 11.1. The labels on the left-hand side of the tree are the splitting criteria or the dimensions to be split.

A *k-d* tree reduces the search cost of a list of N data from an average $O(N/2)$ to an average of $O(\ln N)$. For example, for a database of 10,000 images, the average search cost of a *k-d* tree is *integer* [ln (10,000)] $\approx$ 14, which is way smaller than 5,000, which would be needed for an exhaustive search of a data list. For very large image database, more efficient data structures can be used such as *n*-ary *k-d* tree, *quad-tree*, *octree*, *R-tree*, *cluster* tree, etc.

11.2   Inverted File Indexing

The data structures described above are for numerical data. If images in a very large database have been labeled with nominal or discrete values, they are equivalent to structured textual documents as shown in Fig. 11.2 [1]. Therefore, labeled images can be indexed using the same technique used for textual document indexing. In this section, we first review the inverted file for textual documents indexing and then introduce the inverted file for image indexing.

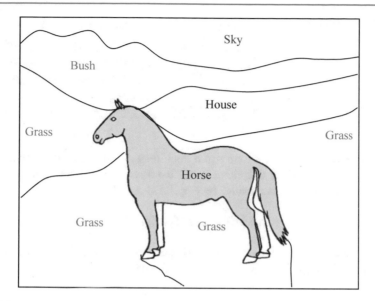

Fig. 11.2 An image with labeled regions

11.2.1   Inverted File for Textual Documents Indexing

In textual document indexing, a technique called *inverted file* is used. An inverted file is a (*terms, docs*) table instead of an ordinary (*docs, terms*) table. The reason of using an inverted file for document indexing is that a large amount of documents (millions/billions, e.g., web pages) can be indexed using a much shorter list of dictionary words (thousands). This makes the search for a large amount of documents very efficient.

The ith entry in an inverted file is a vector: (*term$_i$, doc$_1$, doc$_2$, ..., doc$_n$*), where every document *doc$_j$* has *term$_i$*. Because a term carries a different amount of information in each of the documents, *term$_i$* is given a different weight *tw$_j$* for each *doc$_j$*. Therefore, an inverted file can be shown in Table 11.1.

Since the documents are typically sorted in descending order of importance according to term weights (*tw$_j$*), the term weights can be omitted after sorting. Therefore, an actual inverted file is as simple as Table 11.2.

Table 11.1 A conceptual inverted file for textual document indexing

Term ID	Terms	Documents (weighted)
1	Apple	(doc_{11}, tw_{11}), (doc_{12}, tw_{12}), (doc_{13}, tw_{13}), ...
2	Computer	(doc_{21}, tw_{21}), (doc_{22}, tw_{22}), (doc_{23}, tw_{23}), ...
3	Tree	(doc_{31}, tw_{31}), (doc_{32}, tw_{32}), (doc_{33}, tw_{33}), ...
...	...	...
n	Zebra	(doc_{n1}, tw_{n1}), (doc_{n2}, tw_{n2}), (doc_{n3}, tw_{n3}), ...

Term ID	Terms	Documents (ranked)
1	Apple	$doc_{11}, doc_{12}, doc_{13}, ..., doc_{1n}, ...$
2	Computer	$doc_{21}, doc_{22}, doc_{23}, ..., doc_{2n}, ...$
3	Tree	$doc_{31}, doc_{32}, doc_{33}, ..., doc_{3n}, ...$
...	...	...
n	Zebra	$doc_{n1}, doc_{n2}, doc_{n3}, ..., doc_{nn}, ...$

Table 11.2 An inverted file after sorting

The *term weight* (*tw*) of a term t in a document d is determined by two factors: the *term frequency* of t in document d and the *inverse document frequency* of t in the entire database D. Specifically, *tw* is defined as follows:

$$tw(t,d) = tf(t,d) \times idf(t,D) \tag{11.1}$$

where *tf* stands for *term frequency* and *idf* stands for *inverse document frequency*. The *tf* and *idf* are given in (11.2) and (11.3), respectively.

$$tf(t,d) = \frac{f(t,d)}{\sum_{t_i \in d} f(t_i,d)} \tag{11.2}$$

$$idf(t,D) = \log\left(\frac{\|D\|}{df(t)}\right) \tag{11.3}$$

where

- $f(t, d)$ is the number of occurrence of term t in document d;
- $df(t)$ stands for the *document frequency* of t, it is the number of documents in D which have term t;
- $\|D\|$ is the total number of documents in D.

Recent research shows that the location where a term appears in a document is also a factor in determining the term weight, e.g., a term t in the *title* or *head* section of a web document is given a much higher score because the information of a term in a title or a head is much more important than that in the *body* text.

11.2.2   Inverted File for Image Indexing

The inverted file indexing method can also be applied to image indexing. Once the regions in an image database have been labeled with semantic concepts, images in the database are essentially translated into textual documents. Therefore, images can now be indexed and retrieved the same way as textual documents. Specifically, images in the database are indexed using an inverted file structure, where each index is a vector of the form: (*term*, *image*1, *image*2, ...).

Since each term of an image is associated with a number of image regions, the term weight of an image term is determined by three factors: *area*, *position*, and spatial *relationship* [1, 2]. As a result, three corresponding weights have been defined respectively: *aw*, *pw*, and *rw*.

11.2.2.1   Determine the Area Weight *aw*

Let

- $aw(t)$ be the *area weight* of term t in image I,
- $R(t)$ be the area of a region which is labeled with term t in image I.

Then, $aw(t)$ is defined as the sum of all $R(t)$ in image I normalized by the area of the image. $aw(t)$ is equivalent to the term frequency in the textual document indexing. Mathematically,

$$aw(t) = \frac{\sum_{R\in I} R(t)}{\|I\|} \tag{11.4}$$

For example, in Fig. 11.3 [2], the 4 pink and reddish regions in the center of the image are all labeled as a term of "flower" by a classifier, therefore, the weight of the term "flower" is determined by the total area of the 4 regions, which is the area of the single flower in the image.

11.2.2.2   Determine the Position Weight *pw*

It is known that each type of objects usually has its natural position in an image, e.g., sky is naturally located at the top, grass is naturally located at the bottom, animals are naturally located at the center, etc. Based on this observation, a position weight can be defined for each term of an image document.

Fig. 11.3 A flower in the center with 4 segmented regions

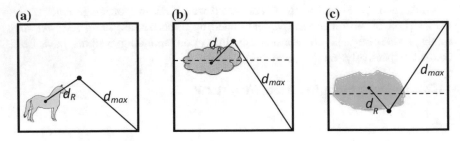

Fig. 11.4 Calculation of d and d_{max} for **a** animal, **b** sky and **c** grass regions

Let

- $R(t)$ be one of the regions in image I associated with term t;
- d_R be the distance between the centroid of the region $R(t)$ and the center of $R(t)$'s natural position in image I;
- d_{max} be the maximum distance between the center of $R(t)$'s natural position and the boundary of image I.

Then, the *position weight* of term t is defined as

$$pw(t) = \sum_{R(t)\in I} 2 \times \left(1 - \frac{d_R}{d_{max}}\right) \tag{11.5}$$

Figure 11.4 shows three examples of computing d_R and d_{max} [1, 2].

11.2.2.3   Determine the Relationship Weight *rw*

It is found that many types of objects usually go together in images such as bird and sky, computer and desk, beach and water, mammals and grass, flowers and tree, car and road, clock and wall, window and building, etc. The *co-occurrence* relationship can be used to determine the weight on a term in an image. For example, if both a "bird" term and a "sky" term are detected in an image, the weight of the "bird" term is doubled; if a "kangaroo" term appears together with a "tree" term and a "grass" term, the weight of the "kangaroo" term is tripled, so on so forth.

Let $r(t)$ be a term which co-occurs with term t in the image I, then the relationship weight *rw* is given as

$$rw(t) = \sum_{r(t)\in I} r(t) \tag{11.6}$$

Now that we have defined the three factor weights of a term t in image I, the final term weight is defined as follows (11.7):

$$tw(t) = aw(t) \times pw(t) \times rw(t) \tag{11.7}$$

Table 11.3 A conceptual inverted file for image document indexing

Term ID	Terms	Images (weighted)
1	Apple	(im_{11}, tw_{11}), (im_{12}, tw_{12}), (im_{13}, tw_{13}), ...
2	Computer	(im_{21}, tw_{21}), (im_{22}, tw_{22}), (im_{23}, tw_{23}), ...
3	Tree	(im_{31}, tw_{31}), (im_{32}, tw_{32}), (im_{33}, tw_{33}), ...
...	...	...
n	Zebra	(im_{n1}, tw_{n1}), (im_{n2}, tw_{n2}), (im_{n3}, tw_{n3}), ...

Table 11.4 The simplified inverted file from Table 11.3 after sorting

Term ID	Terms	Documents (ranked)
1	Apple	$im_{11}, im_{12}, im_{13}, ..., im_{1n}$, ...
2	Computer	$im_{21}, im_{22}, im_{23}, ..., im_{2n}$, ...
3	Tree	$im_{31}, im_{32}, im_{33}, ..., im_{3n}$, ...
...	...	...
n	Zebra	$im_{n1}, im_{n2}, im_{n3}, ..., im_{nn}$, ...

11.2.2.4   Inverted File for Image Indexing

Since each term in the dictionary has been given a weight in each of the images in the database, images in the database can be indexed using an inverted file the same way as in the textual document indexing. An example of an inverted file for image indexing is shown in Table 11.3.

After sorting the images at each row in descending order of importance according to the term weights, the above-inverted file is simplified as Table 11.4. The key difference between Tables 11.4 and 11.2 is that the terms in Table 11.4 are extracted from content features and by machine instead of interpretations by humans. Compared with textual documents, it's more difficult to determine the weight of a term in an image. We will show how the inverted file is used in image retrieval in Sect. 13.5.

11.3   Summary

Image indexing is about to put images in an image database into a data structure or order so that images in the database can be retrieved similar to retrieving alphabetic data from a Relational Database Management System (RDBMS). There are generally two types of approaches: numerical indexing and inverted file indexing.

If images are represented in numerical features, they can be indexed either using a list which is the simplest or using a tree structure. The list indexing is suitable for a small image database, while for a very large image database, the tree structure facilitates fast searching.

If images in a database have been semantically labeled using either machine learning or manual annotation, they can be indexed using an inverted file similar to textual documents indexing. The image database is then translated to an RDBMS. However, the difficulty lies in the determination of term weight. The chapter demonstrates a method of determining term weight using regional features of an image.

For large image databases, both numerical and inverted file indexing are necessary. As shown in Chap. 13, numerical indexing is typically used for query by example while inverted file indexing is used for query by keywords.

References

1. Islam M (2009) SIRBOT—semantic image retrieval based on object translation. PhD thesis, Monash University
2. Zhang D, Islam M, Lu G (2013) Structural image retrieval using automatic image annotation and region based inverted file. J Vis Commun Image Represent 24(7):1087–1098

Image Ranking

12

All that glitters is not gold.

12.1 Introduction

The image feature extracted is usually an N-dimensional feature vector which can be regarded as a point in R^N space. Once images are indexed into the database using the extracted feature vectors, the retrieval of images is essentially the determination of similarity between a query image and the target images in database, which in turn is the determination of distance between the feature vectors in R^N space. The desirable distance measure should reflect human perception. That is to say, perceptually similar images should have smaller distance between them while perceptually different images should have larger distance between them.

Therefore, given a query, the higher the retrieval accuracy, the better the distance measure. For online retrieval, computation efficiency is also a factor to be considered when choosing a distance measure.

Variety of distance measures have been used in image retrieval; they include city block distance, Euclidean distance, cosine distance, histogram intersection distance, χ^2 statistics distance, quadratic distance, and Mahalanobis distance [1]. In this chapter, commonly used similarity measures will be described and examined. A number of widely used performance measurements will also be discussed.

12.2 Similarity Measures

12.2.1 Distance Metric

A similarity measure $d(\mathbf{x}, \mathbf{y})$ between two feature vectors $\mathbf{x}$ and $\mathbf{y}$ is normally defined as a metric distance. $d(\mathbf{x}, \mathbf{y})$ is a metric distance if for any of two data points $\mathbf{x}$ and $\mathbf{y}$ in space; it satisfies the following properties:

© Springer Nature Switzerland AG 2019
D. Zhang, *Fundamentals of Image Data Mining*, Texts in Computer
Science, https://doi.org/10.1007/978-3-030-17989-2_12

(1) $d(\mathbf{x}, \mathbf{y}) \geq 0$ (non-negativity)
(2) $d(\mathbf{x}, \mathbf{y}) = 0$ if and only if $\mathbf{x} = \mathbf{y}$ (identity)
(3) $d(\mathbf{x}, \mathbf{y}) = d(\mathbf{y}, \mathbf{x})$ (symmetry)
(4) $d(\mathbf{x}, \mathbf{z}) \leq d(\mathbf{x}, \mathbf{y}) + d(\mathbf{y}, \mathbf{z})$ (triangle inequality).

12.2.2   Minkowski-Form Distance

The Minkowski-form distance is often called the L_p norm or L_p distance. Given a N-dimensional feature vector of a query image $\mathbf{x} = (x_1, x_2, \ldots x_n)$ and a target image in database $\mathbf{y} = (y_1, y_2, \ldots, y_n)$, the L_p distance is defined as

$$L_p(\mathbf{x}, \mathbf{y}) = \left(\sum_{i=1}^{n} (x_i - y_i)^p \right)^{\frac{1}{p}} \tag{12.1}$$

When $p = 1$, L_1 is called the *city block distance* or *Manhattan distance*:

$$L_1(\mathbf{x}, \mathbf{y}) = \sum_{i=1}^{n} |x_i - y_i| \tag{12.2}$$

When $p = 2$, L_2 is called the *Euclidean distance*:

$$L_2(\mathbf{x}, \mathbf{y}) = \sum_{i=1}^{n} (x_i - y_i)^2 \tag{12.3}$$

When $p \to \infty$, L_∞ is called the *Chebyshev distance*:

$$L_\infty(\mathbf{x}, \mathbf{y}) = \max_{1 \leq i \leq n} \{|x_i - y_i|\} \tag{12.4}$$

By varying the p values, various Minkowski distances can be created. However, among the many Minkowski-form distances, L_2 is the most widely used similarity measures. This is because L_2 is the most consistent with human perception of image similarity. The agreement between distance and perception is demonstrated in Fig. 12.1, where the unit circles of Minkowski distance with different p values are shown. Points on each of the unit circles all have the same distance to the origin under the corresponding p values. As can be seen, the L_2 unit circle agrees most with human perception among the three p values.

L_2 tends to emphasize or amplify the dimensions with high values due to the use of quadratic function. This can cause undesirable results because the distance value is often determined by a few dominant feature dimensions which are often due to local distortion or noise. This in turn can result in rejecting true positives which are perceptually similar images to the query but have local distortion or noise, e.g., a bite out apple would be rejected from the retrieval list using an intact apple as the

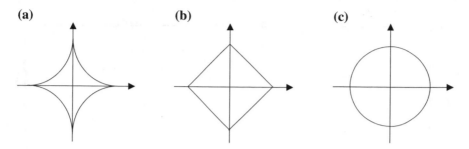

Fig. 12.1 Unit circles of Minkowski distance with different p values. **a** $p = \frac{1}{2}$; **b** $p = 1$; **c** $p = 2$

query. Consequently, L_2 distance can expect lower recall compared with L_1 distance although it can return a top retrieval list with higher precision.

One solution to overcome the lower recall issues of L_2 distance is to apply a logarithm transform to the feature values to suppress the very high feature values and raise the lower feature values, so that all feature values have balanced contributions to the final distance value. Figure 12.2 (top) shows an example histogram from the flower image in Fig. 4.14, notice the histogram feature is dominated by the bins at the end of the histogram. The log-transformed histogram feature vector is shown at the bottom of the figure; it can be seen that while the difference between the feature dimensions has been reduced significantly, the top profile of the histogram has been kept.

12.2.3 Mass-Based Distance

Minkowski-form distance-based similarity measures are basically a matching of two images feature by feature. However, due to image features usually have very high dimensions and features are imperfect, this kind of detailed feature by feature matching can result in undesirable outcomes in many situations. For example, different images can have the same feature vector as shown in Fig. 12.3, and similar images can also have almost completely different feature vectors as shown in Fig. 12.4. In both cases, the L_p-based similarity measure would give a totally incorrect matching result.

The issue demonstrates L_p-based similarity measures that are not robust. This drawback can be overcome by incorporating neighboring data in the decision-making process.

To address the sensitivity issue of L_p, a mass-based similarity measure m_p has been proposed [2]. The idea of m_p is to use neighborhood data to make a similarity decision collectively instead of making a similarity decision just based on two instances alone. Specifically, m_p uses the neighborhood *data mass* at each subspace of R^d to replace the *difference* at each dimension in the Minkowski-form distance.

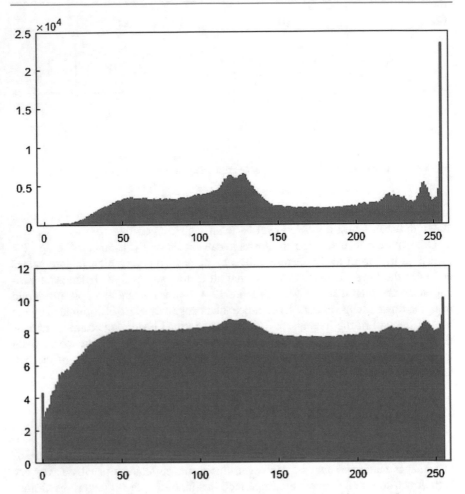

Fig. 12.2 Top: a histogram feature vector; Bottom: the log-transformed histogram feature vector from the top histogram

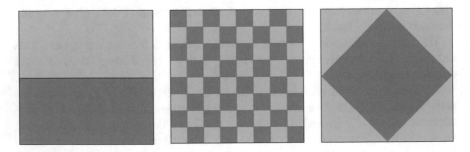

Fig. 12.3 The three images have the same histogram

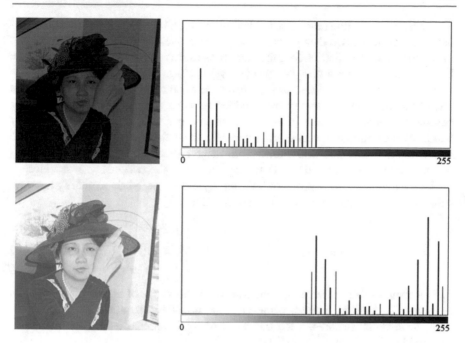

Fig. 12.4 The two images with different brightness have almost completely different histograms

The idea of m_p is based on a distance–density model described by Krumhausl [3] and a psychological discovery that two instances in a sparse region are perceptually more similar than they are in a dense region.

Given two data points in R^n: $\mathbf{x}$ and $\mathbf{y}$, m_p works by defining a region $R(\mathbf{x}, \mathbf{y})$ between the two instances (including the two instances) and finding the data mass of the region. Data mass is the number of data instances from dataset that falls in this region. $R(\mathbf{x}, \mathbf{y})$ is a d-dimensional region, and the ith dimension of $R(\mathbf{x}, \mathbf{y})$ is given as $R_i(\mathbf{x}, \mathbf{y})$, $i = 1, 2, \ldots, n$. The data mass of each $R_i(\mathbf{x}, \mathbf{y})$ depends on the distribution of the data in R^n space.

Specifically, the mass-based similarity measure m_p is defined as (12.5) [4]

$$m_p(\mathbf{x}, \mathbf{y}) = \left(\frac{1}{n} \sum_{i=1}^{n} \left(\frac{|R_i(\mathbf{x}, \mathbf{y})|}{N} \right)^p \right)^{1/p} \tag{12.5}$$

where

- $|R_i(\mathbf{x}, \mathbf{y})|$ is the data mass in region of $R_i(\mathbf{x}, \mathbf{y})$,
- N is the total number of instances in the dataset,
- $R_i(\mathbf{x}, \mathbf{y}) = [\min(x_i, y_i) - \sigma, \max(x_i, y_i) + \sigma]$, and
- σ is a small number and $\sigma \geq 0$.

Figure 12.5 [4] illustrates a data distribution in 2D space and the calculation of data mass between two data points $\mathbf{x}$ and $\mathbf{y}$. For convenience of calculation, σ is set as 0. With this data distribution, the data mass in $R_1(\mathbf{x}, \mathbf{y}) = [x_1, y_1]$ is $|R_1(\mathbf{x},\mathbf{y})| = 63$ while the data mass in $R_2(\mathbf{x}, \mathbf{y}) = [x_2, y_2]$ is $|R_2(\mathbf{x},\mathbf{y})| = 40$.

L_p is essentially a *fine similarity measure* between two instances and is sensitive due to the use of feature by feature matching between two instances. It can result in completely incorrect match in cases shown in Figs. 12.2 and 12.3. On the other hand, m_p is essentially a *coarse similarity measure* between two instances, because it is computed using collective info from neighborhood data mass. Therefore, m_p can be *inaccurate* in situations when the features of the two instances are close.

To overcome the limitations of both the L_p and m_p, a hybrid similarity measure called h_p can be used, which is defined in (12.6)

$$h_p(\mathbf{x},\mathbf{y}) = \left(\sum_{i=1}^{n} (|x_i - y_i| \times |R_i(x,y)|)^p \right)^{\frac{1}{p}} \tag{12.6}$$

h_p is a compromise, it overcomes the sensitivity drawback of L_p while preserves its accuracy. To prevent h_p from being disproportionally determined by a few dominant dimensional features, a log transform on m_p is applied before computing h_p. The modified h_p is given as (12.7)

$$h'_p(\mathbf{x},\mathbf{y}) = \left(\sum_{i=1}^{n} [|x_i - y_i| \times \log(|R_i(x,y)|)]^p \right)^{\frac{1}{p}} \tag{12.7}$$

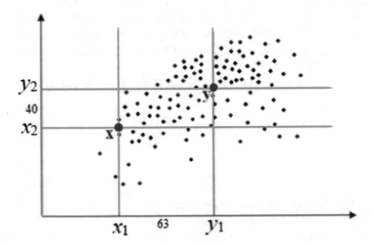

Fig. 12.5 Illustration of m_p dimension calculation between two data points $\mathbf{x}$ and $\mathbf{y}$

12.2.4 Cosine Distance

The *cosine distance* computes the distance between two vectors in terms of direction, irrespective of vector lengths. The distance is computed based on the rule of dot product:

$$\mathbf{x} \times \mathbf{y} = |\mathbf{x}| \times |\mathbf{y}| \times \cos\theta \tag{12.8}$$

where θ is the angle between vector $\mathbf{x}$ and $\mathbf{y}$, and $|\mathbf{x}|$ and $|\mathbf{y}|$ are the magnitudes of $\mathbf{x}$ and $\mathbf{y}$, respectively. The cosine distance is then defined as

$$\cos(\mathbf{x}, \mathbf{y}) = 1 - \frac{\mathbf{x} \cdot \mathbf{y}}{|\mathbf{x}| \cdot |\mathbf{y}|} = \frac{\sum_{i=1}^{n} x_i y_i}{\sqrt{\sum_{i=1}^{n} x_i^2} \sqrt{\sum_{i=1}^{n} y_i^2}} \tag{12.9}$$

If both x_i and y_i have been normalized to probability values between 0 and 1, cos $(\mathbf{x}, \mathbf{y})$ becomes

$$\cos(\mathbf{x}, \mathbf{y}) = 1 - \sum_{i=1}^{n} x_i y_i \tag{12.10}$$

The key feature of the cosine distance is that it is invariant to scale change in contrast to Minkowski distance. Figure 12.6 shows the comparison between the cosine distance and the two Minkowski-form distances in two-dimensional space. It can be observed that both L_2 and L_1 respond to scale changes, while cosine distance does not. For example, in Fig. 12.6b, $\cos(\mathbf{x}, \mathbf{y}) = \cos(\mathbf{x}_1, \mathbf{y})$, while $L_1(\mathbf{x}, \mathbf{y}) \neq L_1$ $(\mathbf{x}_1, \mathbf{y})$ and $L_2(\mathbf{x}, \mathbf{y}) \neq L_2(\mathbf{x}_1, \mathbf{y})$. The scale invariance can be useful in situations where directional features are more important than magnitudes. For example, if cosine distance is used, two similar colors will keep their similarity after scaling of the color components.

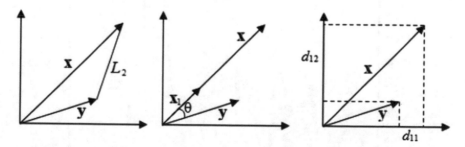

Fig. 12.6 Comparison between the cosine distance and L_p distance. **a** $L_2(\mathbf{x}, \mathbf{y}) = L_2$; **b** $\cos(\mathbf{x}, \mathbf{y}) = \cos\theta$; **c** $L_1(\mathbf{x}, \mathbf{y}) = d_{11} + d_{12}$

12.2.5 χ^2 Statistics

In χ^2 test, both **x** and **y** are treated as random variables, the χ^2 statistics is then used to test if the two variables are correlated/independent each other, and how much they are correlated. Formally, χ^2 statistics is defined as (12.10)

$$\chi^2(\mathbf{x}, \mathbf{y}) = \sum_{i=1}^{n} \frac{(x_i - m_i)^2}{m_i} \qquad (12.11)$$

where $m_i = (x_i + y_i)/2$, which is regarded as the *expected value* for dimension i. A low χ^2 value means that both **x** and **y** are from the same probability distribution and there is a high correlation between the two feature vectors, which indicates the images represented by the two feature vectors are similar. An advantage of using χ^2 statistics is that it can overcome the mismatch between two histograms from images with very different lighting conditions as shown in Fig. 12.4.

12.2.6 Histogram Intersection

A histogram is a distribution function with a particular shape of area. The histogram intersection is to test how much area two distributions **x** and **y** share, the more area they share, the more similar the two distributions are (Fig. 12.7). Mathematically, a histogram intersection is defined as

$$HI(\mathbf{x}, \mathbf{y}) = \frac{\sum_{i=1}^{n} \min(x_i, y_i)}{\min(|\mathbf{x}|, |\mathbf{y}|)} \qquad (12.12)$$

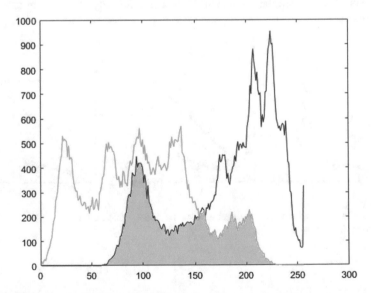

Fig. 12.7 Histogram intersection of two histograms shown as gray area

If both x_i and y_i have been normalized to probability values between 0 and 1, *HI* is simplified as (12.13)

$$HI(\mathbf{x}, \mathbf{y}) = \sum_{i=1}^{n} \min(x_i, y_i) \tag{12.13}$$

For two identical histograms, their *HI* value is the maximum 1 and for two similar histograms, their *HI* value is a high. For two different histograms such as the two histograms shown in Fig. 12.4, their *HI* value is close to zero. The *HI* distance is defined as

$$d_{HI}(\mathbf{x}, \mathbf{y}) = 1 - \sum_{i=1}^{n} \min(x_i, y_i) \tag{12.14}$$

d_{HI} also has the same histogram mismatching issue as the L_p distance.

12.2.7 Quadratic Distance

The distances or measures we have introduced so far all make two implicit assumptions: (a) the two feature vectors to be measured $\mathbf{x}$ and $\mathbf{y}$ have equal number of dimensions; and (b) the dimensions of $\mathbf{x}$ and $\mathbf{y}$ are independent. However, there are applications and situations where these two conditions are not met. For example, the dominant color descriptors described in Chap. 4 typically have different number of dimensions, and colors of neighboring histogram bins are correlated with each other. The quadratic distance measure is one of the methods to address the unequal number of dimensions between two feature vectors and capture the cross dimension information in a feature vector.

The quadratic-form distance between two n-dimensional feature vectors $\mathbf{x}$ and $\mathbf{y}$ is given by

$$d_q(\mathbf{x}, \mathbf{y}) = \left[(\mathbf{x} - \mathbf{y})^T \mathbf{A} (\mathbf{x} - \mathbf{y}) \right]^{\frac{1}{2}} \tag{12.15}$$

where

- T means transpose,
- $\mathbf{A} = [a_{ij}]$ is an $n \times n$ matrix,
- a_{ij} is the similarity coefficient between dimensions i and j,
- $a_{ij} = 1 - d_{ij}/d_{max}$,
- $d_{ij} = |x_i - y_j|$, and
- $d_{max} = \max_{1 \le i,j \le n} d_{ij}$.

For numerical calculations, (12.15) is expanded as (12.16)

$$d_q = \left(\sum_{i=1}^{n} \sum_{j=1}^{n} a_{ij} x_i x_j + \sum_{i=1}^{n} \sum_{j=1}^{n} a_{ij} y_i y_j - 2 \sum_{i=1}^{n} \sum_{j=1}^{n} a_{ij} x_i y_j \right)^{\frac{1}{2}} \quad (12.16)$$

The a_{ij} is the similarity coefficient between x_i and y_j; it is a *weight* on a cross-dimensional element of the two feature vectors, the higher the correlation between the two cross dimensions, the more the weight is given on that element.

For two feature vectors **x** and **y** with different dimensions n and m, respectively, the quadratic distance between **x** and **y** is given as (12.17)

$$d_q = \left(\sum_{i=1}^{n} \sum_{j=1}^{n} a_{ij} x_i x_j + \sum_{i=1}^{m} \sum_{j=1}^{m} a_{ij} y_i y_j - 2 \sum_{i=1}^{n} \sum_{j=1}^{m} a_{ij} x_i y_j \right)^{\frac{1}{2}} \quad (12.17)$$

If the dimensions of both the two feature vectors **x** and **y** are independent each other, e.g., after certain decorrelation operations, the quadratic distance between **x** and **y** is given as (12.18)

$$d_q = \left(\sum_{i=1}^{n} x_i^2 + \sum_{j=1}^{m} y_j^2 - 2 \sum_{i=1}^{n} \sum_{j=1}^{m} a_{ij} x_i y_j \right)^{\frac{1}{2}}$$

$$= \left(\sum_{i=1}^{n} \sum_{j=1}^{m} a_{ij} \left(x_i - y_j \right)^2 \right)^{\frac{1}{2}} \quad (12.18)$$

Equation (12.18) is a *weighted Euclidean distance*; one can expect that d_q is a more desirable similarity measure than both L_2 and d_{HI}; however, the determination of the weights is an issue.

12.2.8   Mahalanobis Distance

The *Mahalanobis distance* is a special case of the quadratic-form distance (12.15) in which the transform matrix is determined by the *covariance matrix* obtained from a training set of feature vectors, that is, $\mathbf{A} = \Sigma^{-1}$. In order to apply the Mahalanobis distance, a feature vector **x** is regarded as a multivariate random variable $\mathbf{x} = (x_1, x_2, \ldots, x_n)$ from certain probability distribution. Then, the correlation matrix is given by **R** where

- $\mathbf{R} = [r_{ij}]$
- $r_{ij} = E\{x_i x_j\}$ which is the mean of the random variable $x_i x_j$.
- The covariance matrix Σ is given by $\Sigma = [\sigma_{ij}^2]$.
- where $\sigma_{ij}^2 = r_{ij} - E\{x_i\} E\{x_j\}$.

The Mahalanobis distance between two feature vectors $\mathbf{x}$ and $\mathbf{y}$ is given as

$$d_m(\mathbf{x}, \mathbf{y}) = [(\mathbf{x} - \mathbf{y})\Sigma^{-1}(\mathbf{x} - \mathbf{y})]^{\frac{1}{2}} \tag{12.19}$$

In the special case where x_i are statistically independent but have unequal variances, Σ is a diagonal matrix as follows:

$$\Sigma = \begin{bmatrix} \sigma_1^2 & & & 0 \\ & \sigma_2^2 & & \\ & & \ddots & \\ 0 & & & \sigma_n^2 \end{bmatrix} \tag{12.20}$$

In this case, the Mahalanobis distance is reduced to a simpler form:

$$d_m(\mathbf{x}, \mathbf{y}) = \left(\sum_{i=1}^{n} \frac{(x_i - y_i)^2}{\sigma_i^2} \right)^{\frac{1}{2}} \tag{12.21}$$

Equation (12.21) is another *weighted Euclidean distance*. It gives more weight to dimensions with smaller variance and less weight to dimensions with larger variance. d_m can be regarded as a *standard Euclidean distance*. The Euclidean distance is just a special case of Mahalanobis distance when the covariance matrix Σ is the identity matrix.

12.3 Performance Measures

After image ranking, we need a measure to tell how good is the ranking by a similarity measure we have discussed above. Specifically, we need to assess how many relevant images have been retrieved on the top list and how many relevant images have missed from the top list. The information from the top list of retrieval lets us tell how well a similarity measure performs. A performance measure is usually based on statistics of a *subjective test* which is a test of identifying relevant images to the query and how relevant they are to the query. Different performance measures often use different subjective tests, resulting in different definitions of retrieval performance. In this section, several commonly used performance measures are described and discussed.

12.3.1 Recall and Precision Pair (RPP)

RPP is one of the most widely used retrieval performance measurements in literature. In RPP, for *each query image*, images in a dataset are divided into two categories: *relevant* images (1) and *irrelevant* images (0), based on their similarity to the query. The similarity is determined by a subjective test on a group of subjects. In the subjective test, each subject selects items relevant to the query from the dataset. An item selected by more than a number of subjects as a relevant image is given a label of "1"; otherwise, it is regarded as an irrelevant image and is given a label of "0".

Now given a query image I and a retrieval list returned by a similarity measure, the *precision* (P) and *recall* (R) statistics are then computed based on the "0" and "1" images presented on the top retrieval list:

$$P = \frac{r}{n_1} = \frac{number\ of\ relevant retrieved\ images}{number\ of\ retrieved\ images}$$
$$= \frac{|\{relevant\ images\} \cap \{retrieved\ images\}|}{|\{retrieved\ images\}|} \tag{12.22}$$

$$R = \frac{r}{n_2} = \frac{number\ of\ relevant\ retrieved\ images}{number\ of\ relevant\ images\ in\ DB}$$
$$= \frac{|\{relevant\ images\} \cap \{retrieved\ images\}|}{|\{relevant\ images\}|} \tag{12.23}$$

P can be interpreted as the probability that a retrieved image is relevant, while R can be interpreted as the probability that a relevant image is returned by a retrieval.

The RPP is often given in the following form:

$$P = \frac{t}{t + f_p} \tag{12.24}$$

$$R = \frac{t}{t + f_n} \tag{12.25}$$

where t, f_p, and f_n stand for "True Positive" (a hit), "False Positive" (a mismatch), and "False Negative" (a miss), respectively.

Precision measures the retrieval accuracy while recall measures the retrieval robustness; both are important for a similarity measure. Precision and recall are inversely related, i.e., precision normally degenerates as recall increases. Translating into actual image retrieval result, this inverse relationship means that the shorter the retrieval list (low recall), the higher the accuracy and vice versa.

The RPP based on a single query does not provide a complete picture of the performance of a similarity measure; usually, a number of queries are tested and the P values at each of the R values are averaged. The average (R, P) values are then plotted on a graph to get an approximate performance of a similarity measure.

Figure 12.8 shows an RPP curve from an averaged retrieval result. It can be observed from the figure that, as the *recall* increases (longer retrieval list), the *precision* goes down rather sharply in this case.

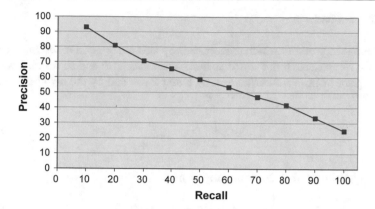

Fig. 12.8 An RPP curve from an actual image retrieval

The RPP curve gives a good picture of a similarity measure's performance. A good similarity measure will have an RPP curve with two characteristics: (a) a high start (how high depends on applications), e.g., 70%+; and (b) a gentle drop. However, it is often difficult to achieve both the two goals; a retrieval method usually either targets high precision on the top list of a retrieval result or targets higher recall depending on applications. Therefore, the P values at the lower recall values are much more important than those at higher recalls. For example, in Fig. 12.8, the P values before the 30% of recall are all above 70%, which indicates a good retrieval result although the full RPP curve does not look good.

Although RPP is intuitive, there are several drawbacks to this performance measure.

- **Need a ground truth**. In order to compute the R value, we need to know the total number of relevant images in a database which is essentially a ground truth. This limits the application of the RPP to databases with small scale.
- **Unrealistic relevance values**. The binary relevance value given to each of the images in the database is not realistic, because image similarity is probabilistic and between 0 and 1.
- **Missing ranking information**. All relevant images on the retrieval list are given the same relevance value; ranking information is not considered in defining the relevance values. However, a similar image at rank 1 is more relevant than a similar image at rank 10.
- **A pair of conflict values**. It is often awkward to tell the performance of a retrieval using *two* values which do not agree with each other. For example, if we have a retrieval which gives $P = 90\%$ and $R = 10\%$, it is difficult to tell how well is the retrieval result. Therefore, we need a measure to reconcile the P and R pair.

A number of other performance measures have been designed to address the above issues.

12.3.2   *F*-Measure

A performance measure which reconciles the *precision* (P) and *recall* (R) into one is called the *F*-measure. It is defined as *harmonic mean* of P and R, which turns out to be the square of the *geometric mean* of P and R divided by the *arithmetic mean* of P and R.

$$F = \frac{P \cdot R}{\frac{P+R}{2}} = 2 \cdot \frac{P \cdot R}{P+R} \tag{12.26}$$

It can be shown that the following is true:

$$F = \alpha P + (1 - \alpha)R \tag{12.27}$$

where $\alpha = \frac{t+f_p}{2t+f_p+f_n}$. Therefore, F turns out to be a weighted sum of P and R. The weight α can be adjusted to suit a specific data or application.

Figure 12.9 shows the F curve against the same P-R curve from Fig. 12.8. It can be observed that the F score has a low start and reaches the maximum value at the point where the P score is the closest to the R score. Before the peak point, precision is more important; after the peak point, recall is more important. Therefore, the peak point is the optimal tradeoff between P and R. Overall, the higher the F score, the better tradeoff between P and R. Therefore, for two similarity measures or retrieval results, the one gives a higher F score is usually better.

The advantage of using F-measure is that a single value can be used to tell the difference between two similarity measures or two retrieval results. However, it is not as intuitive as the RPP and it is not easy to interpret an F score. It appears we could have used an AUC or *area under* (the RPP) *curve*, as an alternative to F-measure. The AUC would be not only a single value but also as intuitive as the RPP. However, the AUC would not be able to differentiate an RPP with high start but sharp drop (a sliding RPP) and an RPP with low start but relatively flat (a steady

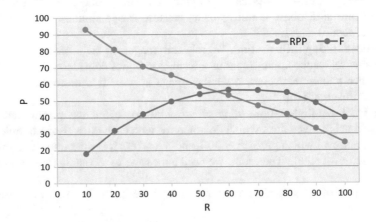

Fig. 12.9 The *RPP* curve and *F* curve from an actual image retrieval

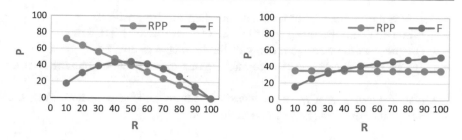

Fig. 12.10 The F curves for two different RPP curves with the same AUC

RPP). The former RPP is usually more preferable than the later one, even though the former is not a good RPP either. Therefore, the AUC would not be as effective as the F-measure. For example, Fig. 12.10 shows a *sliding* RPP and a *steady* RPP with the same AUC. However, the maximum/optimal F score of the slide RPP is at middle of the RPP curve, while the maximum/optimal F score of the steady RPP is at the very end of the RPP curve, which is undesirable, because it is unlikely that a user would wait until all the relevant images are shown up.

12.3.3 Percentage of Weighted Hits (PWH)

The PWH can be regarded as a weighted recall. The subjective test is the same as in RPP, that is, each subject select items relevant to the query from the dataset. However, instead of measuring recall based on *binary relevance value* as in RPP, PWH assigns a *weighted relevance value* w_i to each item in the dataset. The sum of the weights w_i is equivalent to the number of subjects selecting item i as relevant or similar to the query. Therefore, PWH is defined as

$$PWH = \frac{\sum_{i=1}^{n} w_i}{\sum_{i=1}^{N} w_i} \tag{12.26}$$

where n is the number of items retrieved and N is the total number of items in the database. It is easy to see that the R measure in RPP is a special case of PWH when w_i takes the value of 0 and 1. Similar to the R measure, PWH needs to identify every item in the database as relevant or not relevant to the query, and this need for ground truth of the database limits its usage.

12.3.4 Percentage of Similarity Ranking (PSR)

PSR is a performance measure of detecting the agreement between an algorithm ranking and a human ranking [5]. In this method, each subject assigns a similarity rank to each item i in the dataset based on the item's similarity to the query j. For each query j, the final result of the subject test is a matrix $\{Q_j(i, k)\}$, where

- $Q_j(i, k)$ is the number of subjects ranking item i at kth position.
- $\bar{p}_j(i)$ and $\bar{\sigma}_j(i)$ are the mean and variance of each row of $\{Q_j(i, k)\}$.
- $\bar{p}_j(i)$ represents the average ranking of the ith image to query j.
- $\bar{\sigma}_j(i)$ represents the degree of agreement among the subjects on ranking item i.

Given a query j, if a *retrieval algorithm* returns an item i at rank $P_j(i)$, the *percentage similarity ranking* $S_j(i)$ is defined as

$$
S_j(i) = \sum_{k=P_j(i)-\frac{\sigma_j(i)}{2}}^{P_j(i)+\frac{\sigma_j(i)}{2}} Q_j(i, k) \tag{12.27}
$$

A plot of $S_j(i)$ as a function of item i shows the retrieval performance of the retrieval algorithm. A high $S_j(i)$ curve indicates a high retrieval accuracy of the algorithm. An average PSR value can also be computed as the overall performance of the retrieval algorithm.

The PSR takes into account the number and agreement of human ranking. However, if for a query, the percentage of humans giving a particular item at particular ranking is high (high degree of agreement on ranking the item), then the variance for the ranking would be small. This would result in unusually low PSR if the retrieval algorithm's ranking differs from the subject mean ranking. On the other hand, if the variance of a ranking is large, then the PSR would be unusually high even if the ranking by the retrieval algorithm differs substantially from the subject mean ranking.

12.3.5 Bullseye Accuracy

A simple Bullseye performance measure (BEP) is called *Precision at K* or P@K, which is defined as the ratio of the "number of relevant images on the top K retrieval" to K. This ratio is called a *Bullseye score*. The higher the Bullseye score, the better the retrieval. P@K is a very convenient and useful performance measure because it does not need the ground truth of the database. P@K measure is widely used in applications where the data is massive and the accuracy of the top retrievals is the most important. For example, in online web search, users only care about the relevance of the top pages returned by a search engine.

The K can be determined based on the actual data or application. If the total number of relevant images in the database is known to be N, K is typically determined as N or $2N$. In practice, Bullseye scores are obtained from a number of queries and an average Bullseye score is obtained as the overall performance value for a retrieval algorithm.

The *Bullseye score* can also be defined as the *Average rank* (AR). In AR, instead of precisions, the ranks of relevant images on the top K retrieval list are averaged. The lower the AR, the better the retrieval result.

12.4 Summary

Once images are stored and indexed, similarity measure and performance measure work together to rank images similar to a query. To draw an analogy, image retrieval is like gold mining. A similarity measure is analogous to a refinery procedure in gold mining and the performance measure is equivalent to a laboratory test which determines if a refinery procedure produces sufficient yield of pure gold.

Design of similarity measures can be divided into three categories: (a) Geometric-based methods such as L_p, cosine distance; (b) Statistical methods such as $\chi 2$ statistics, histogram intersection d_{HI}, and Mahalanobis distance d_m; and (c) Hybrid methods such as h_p, m_p, quadratic distance d_q. Zhang and Lu [1] has made an evaluation of the commonly used similarity measures using the MPEG-7 shape image databases; it has been found that the *city block distance*, *Euclidean distance*, and *$\chi 2$ statistics* are among the top performance similarity measures. The possible reason to explain this finding is that the three distances are all simple. The more complex the similarity measure is, the more unpredictable the result.

The design of performance measures is based on three types of criteria: (a) Accuracy, such as P and P@K; (b) Robustness, such as R and PWH; and (c) Both accuracy and robustness, such as RPP, F, PSR, and AR. Each of these performance measures can be varied by how the relevance value is determined and how the ranking information is used. Overall, the RPP and P@K are the most intuitive and have the least complexity.

Although we have only demonstrated how to use these performance measures to evaluate the performance of the similarity measures introduced in this chapter, they can also be used to evaluate the performance of different image features described in Part II in the same way.

References

1. Zhang D, Lu G (2003) Evaluation of similarity measurement for image retrieval. In: Proceedings of neural networks and signal processing
2. Aryal S et al (2017) Data-dependent dissimilarity measure: an effective alternative to geometric distance measures. In: Knowledge and information systems, pp 1–28
3. Krumhansl C (1978) Concerning the applicability of geometric models to similarity data: the interrelationship between similarity and spatial density
4. Shojanazeri H, Teng S, Aryal S, Zhang D, Lu G (2018) A novel perceptual dissimilarity measure for image retrieval. In: Proceedings of IVCNZ2018
5. Bimbo A, Pala P (1997) Visual image retrieval by elastic matching of user sketches. IEEE Trans PAMI 19(2):121–132

Image Presentation

13

The best usually is also the simplest.

13.1 Introduction

Image presentation is about how present the database images or retrieved images to the user in the most effective and efficient way. Given a query from a user, the list of images retrieved from the database can number from thousands to millions. Due to the limitation of both image features and image ranking, the retrieval list is usually scattered with irrelevant images. How to organize the retrieved images and present as many relevant images as possible in a very limit space is a great challenge in IR. Image presentation is part of the research on user interface and data visualization. It is a mixture of tech and arts. Many interesting image presentation methods have been developed, and they include simple browsing, category browsing, content browsing, hierarchical organization, sophisticated approaches involving user interaction, etc. In this chapter, we discuss a number of most common approaches on image presentation.

13.2 Caption Browsing

If images in the database have been labeled with captions, the simplest way to find relevant images is through caption browsing. This is typically used in personal/family photo albums stored in PCs and photo galleries on web pages, which are usually labeled, dated, and grouped by photo takers.

Caption browsing can also be used in image retrieval. If a retrieval list is not long, this method can be used to scan through the images on the list in order to find the most relevant images. Figure 13.1 shows a caption browsing given by MS Windows File Explorer.

© Springer Nature Switzerland AG 2019

D. Zhang, *Fundamentals of Image Data Mining*, Texts in Computer Science, https://doi.org/10.1007/978-3-030-17989-2_13

Fig. 13.1 Browsing images using image captions by Windows File Explorer

The advantage of using caption browsing is that it is effortless and it suits for universal users. Furthermore, users can search images based on image captions or date information. For small image databases or a short image retrieval list, caption browsing is the best option. A major issue with caption browsing is that the caption labels are subjective, and they are often confusing and even misleading.

13.3   Category Browsing

Category browsing is a typical approach for organizing image databases which are much larger than personal or family photo albums. Suppose images have been classified or annotated using the methods described in Part III; images can be organized into categories just like books are organized in library. Indeed, this is exactly how images are traditionally indexed in library and archives. Under each category, users can further filter the images using a simple browsing. Figure 13.2 shows an example of category browsing used by MS Windows File Explorer.

The advantage of category browsing is that very large collections of images can be broadly organized into categories every efficiently; there is no need to label individual images. This is particularly useful given there is massive amount of digital images on the Internet and most of the images are either unlabeled or mislabeled. However, images in categories of very large image collections are

Fig. 13.2 Category browsing using MS Windows File Explorer

usually too diverse both perceptually and semantically. A common approach is to subdivide each of the categories into subcategories and a more efficient hierarchical category browsing is designed.

13.3.1 Category Browsing on the Web

More effective and efficient category browsing methods have been designed for the web where each category is initiated with one or more representative images. Figure 13.3 shows an example website using category browsing. It can be seen that the Web category browsing is more effective and efficient than a desktop file explorer, because it not only visualizes each category with one or multiple iconic images but also provides a convenient searching mechanism. Furthermore, a web browsing system also provides a user with key information about the image collection and categories, such as the number of categories, number of images, types of images, source of images, instructions on retrieving the images, how to use the images, etc. More examples of category browsing on the web can be found online by Googling for "image categories."

13.3.2 Hierarchical Category Browsing

Categories of the same concept can be merged to organize the image database into a *hierarchical* or *tree* structure. On the other hand, if there are too many images in a category, it can be subdivided into subcategories. The subjective labeling issue in image categorization can be overcome using a thesaurus such as the WordNet [1].

The design of a hierarchical category browsing system can be done either before the collection of images or during the collection of images. Due to the lack of

Fig. 13.3 A website of category browsing

standard image taxonomy, image categories are typically defined during the collection of images or defined "on the run". In this case, images should be put into as many categories as possible, so that higher level of categories can be created by merging similar subcategories using a thesaurus. Images can be most efficiently retrieved using a standard hierarchy like the one used to classify animals or plants.

Figure 13.4 shows an example of hierarchical organization of image categories; each top-level category is demonstrated with a number of iconic images to give an

Fig. 13.4 Hierarchical organization of image categories on a website

idea on what is inside the category. It can be seen that it is a more effective and efficient presentation of image categories than that in Fig. 13.3, because more image categories can be shown to users on one web page.

13.4 Content Browsing

Images in the database or on a retrieval list can also be organized based on *numerical image features* instead of semantic labels as shown in the above sections. Numerical image features are also called low-level image features, perceptual features, or *content features*. Therefore, image retrieval based on numerical image features is called *content-based image retrieval* or CBIR for short. The idea of *content browsing* is to convert and group retrieved images as thumbnails based on their perceptual similarities such as color, texture, or shape. The organization of the thumbnails is then presented as an *image map* to guide a user to navigate to the most relevant images. This is very useful because people usually judge image similarity based on perception, and perceptual features are usually less subjective than semantic labeling which can vary widely from person to person. Figure 13.5 shows an example of content browsing.

Content browsing is a study of image *data visualization*. Key issues in content browsing include the following:

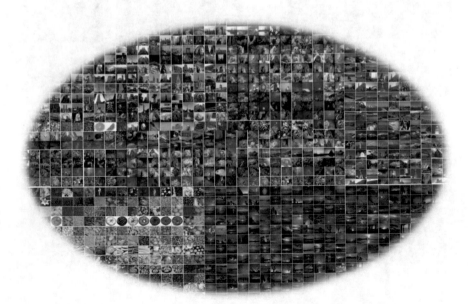

Fig. 13.5 Image presentation based on content browsing

- **Features** of images. What image feature(s) to use? Typically, images to be presented are grouped based on color, texture, shape, or combination of them, and it depends on the nature of the image data and application.
- **Similarity** measure. What similarity measure to use? Different similarity measures can lead to different perceptual experiences on the content browsing.
- **Size** of thumbnails. Should they be equal size or different sizes depending on relevance?
- **Overlap** of thumbnails. Should thumbnails be overlapped and how much?
- **Structure and layout** of thumbnails. Should they be organized in row and column, spiral, or graph?
- **Space** of presentation. Whether images are presented in 2D space or 3D space such as a cylinder or a sphere.
- **Interaction**. What type of interaction to use and how a user can interact with the thumbnails?

These factors are determined by research on how the physical world can be most effectively presented to human vision system.

13.4.1 Content Browsing in 3D Space

The limited space of a 2D screen can be extended using a 3D cylinder (Fig. 13.6) [2], or a sphere (Fig. 13.7) to present more thumbnails in an image map [3]. Images can also be presented on a curved wall as shown in Fig. 13.8 to extend the visual space.

13.4.2 Content Browsing with Focus

Content browsing with equal size thumbnails provides a global view of the visualized image data. Alternatively, a local view of the image map provides a content browsing with *focus*. For example, in Fig. 13.9, images on the top of the retrieval list are given focus and presented in the center of the screen with larger size according to their similarity to the query image, and they are then arranged in a clockwise spiral order [4].

13.4.3 Force-Directed Content Browsing

Conventional content browsing does not have overlap between images and the distance between neighboring images is uniform. In a force-directed content browsing, images in a category or a retrieval list are regarded as planets in space and the distance between images is regarded as a force between them. If two images are similar enough, the force is strong enough to attract them together. The result of a force-directed visualization is an overlapped image map as shown in Fig. 13.10a.

Fig. 13.6 Content browsing using a cylindrical image map

Fig. 13.7 Content browsing
using a spherical image map

Figure 13.10b shows the force-directed image map in diminishing perspective. A bi-force-directed content browsing can be visualized using a 3D image map as shown in Fig. 13.10c. In the 3D image map, there are two types of forces

Fig. 13.8 Content browsing using a curved wall image map

representing two types of similarity between images, e.g., the horizontal force represents the color similarity and the vertical force represents the texture similarity. A close-up section of the 3D image map is shown in Fig. 13.10d.

Interactivity over an image map can also include keywords. For example, keywords can be used as a mouse-over feature of the image map. A *keyword map* and an *image map* can be created separately and put side by side, and the two maps are linked to provide a user with a joint content browsing [4].

13.5 Query by Example

When come to find a required image, one of the most challenging issues for a user is how to initiate the search, which includes how to start, where to start and how to formulate a search. One of the simplest ways to start a search is a *query by example* (QBE), or search based on an example image from the user. In QBE, similar images in the database are retrieved based on the *content features* of the query image. The

Fig. 13.9 Content browsing with focus and spiral structure

justification for QBE is that often the requirement for an image cannot be described by words and a user has a sample image which is similar to what they actually want to find. The sample image can be used as equivalent keywords for a query. QBE is a useful functionality; because most of the images in the world are unlabeled, these images can be retrieved based on content features. Therefore, QBE is an essential part of an IR system.

An example of QBE is Google's "Search by image" as shown in Fig. 13.11 [5]. In the figure, the sample image was first *translated* to the keywords "Sydney opera house" which are then used as the query for the actual image retrieval. Due to the translation, Google's search by an example image can be regarded as a semi-QBE, because labels of images in the database may not be obtained from image content. This works well when a query image is properly translated; however, when a query image is incorrectly translated, it results in completely wrong retrievals. For example, Fig. 13.12 shows three example queries which are translated to "darkness", "comics", and "computer", respectively. As a result, the retrievals of the three queries are completely incorrect. Furthermore, when the hand image query is presented using two different copies, they are translated as "Naya Rivera" and

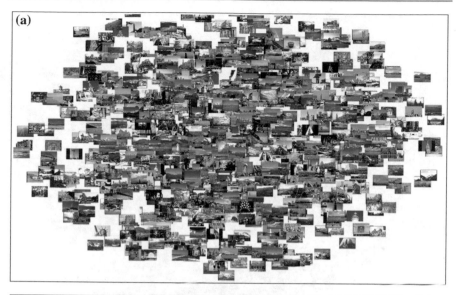

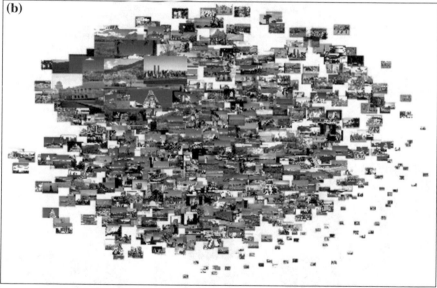

Fig. 13.10 Force-directed content browsing

"Point Cabrillo Light," respectively, which would not occur if content features are used for direct matching. This demonstrates that a CBIR-based QBE is a necessary component for an image retrieval system.

Although QBE is useful and convenient in many situations, it can be challenging to find an example image for a user. One of the ways to find example images is to use *query by keywords* or QBK which will be introduced in the following section.

(c)

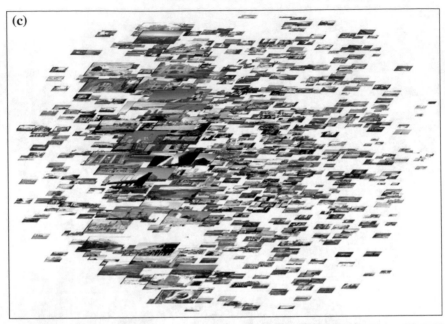

(d)

Fig. 13.10 (continued)

Fig. 13.11 A QBE example from Google

Fig. 13.12 Query images for Google's "Search by image"

13.6 Query by Keywords

People describe the visual world using textual language such as a blue sky, green trees, red flowers, a cheerful street, violent protests, etc. One of the natural ways to search images is to use *query by keywords* or QBK. In a QBK system, images in database are all labeled with keywords and indexed using inverted files; the search for images is just the same as search for textural documents. Unfortunately, most of the images in the world are either unlabeled or mislabeled. In a typical QBK system, image labels are obtained from image content through automatic image classification/annotation. A typical QBK system is shown in Fig. 13.13 [6, 7]. Image labels can also be obtained by analyzing textual description of images in web pages on the Internet.

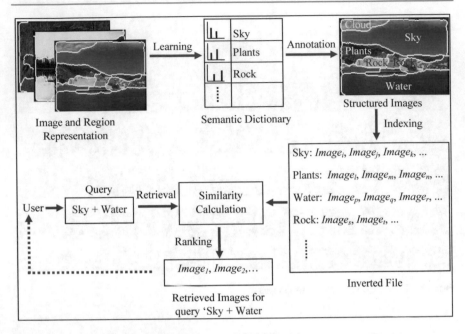

Fig. 13.13 A content-based QBK system

One of the most widely used QBK systems is Google's image search as shown in Fig. 13.14a [5]. The key features of Google's image search are summarized as follows:

- **Simplicity**. The textual input is the simplest and most intuitive query interface. It is a universal user interface.
- **Caption browsing**. Retrieved images are presented as caption browsing which is simple and intuitive.
- **Category browsing**. The categories at the top of the browsing area provide users with category browsing (Fig. 13.14b).
- **Content browsing**. QBE is included as an option for the QBK interface, shown as a camera icon at the query field (Fig. 13.14a). Some of the category browsings are also based on content features such as colors.
- **Interactivity**. An enlarged version of an image is shown as focus when a user clicks any of the images in the browsing area; the enlarged image is also linked to the image source (Fig. 13.15).
- **Hierarchical structure**. When a user clicks an image in the browsing area, it also shows the category of images similar to the clicked image at the right-hand side (Fig. 13.15). This kind of interaction can also be regarded as a QBK-supported QBE.

At this moment, Google's image search is predominantly done by QBK due to the limited capability of CBIR techniques. However, it is often a challenge for a user to formulate an image query using keywords only. In many cases, a query for

(a)

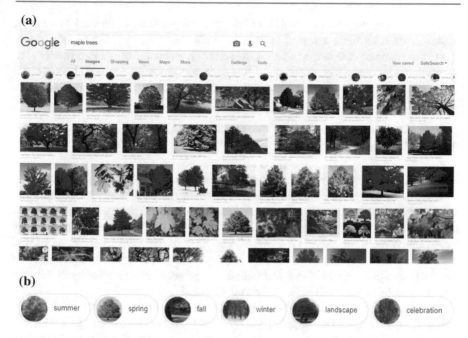

(b)

Fig. 13.14 A QBK example from Google. **a** An example from Google's Image search; **b** category browsing from Google's Image search

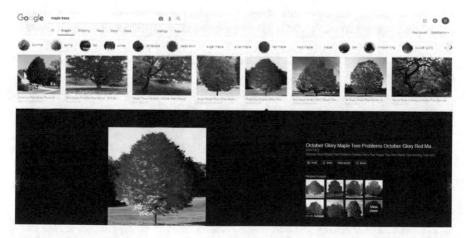

Fig. 13.15 A hierarchical QBK in Google. An enlarged image at the black strip and its relevant category at the right-hand side of the enlarged image

an image can only be described by a visual form or QBE. As CBIR techniques become more mature, it is expected that both QBE and QBK will play equal roles in image search.

An effective *relevance feedback* or RF mechanism is an essential feature of an image retrieval system. This is because an initial retrieval result is likely inaccurate; the RF provides a user a chance to interfere and refine the retrieval result. RF works by letting a user to select a number of relevant images from a retrieval list; the IR system then uses these relevant images as a collective query (instead of a single query) to refine or update the retrieval list so that it meets the user's expectation. RF is particularly useful for mass scale image search such as Google, because the chance of identifying relevant images from the top of a retrieval list is high.

13.7   Summary

In this chapter, different image presentation techniques have been discussed. Once images in an image database have been classified, indexed, and ranked, the next step is how to present the images to users. We have introduced a number of common methods of image presentation including caption browsing, category browsing, content browsing, QBE, QBK, etc. Each of them has its applications and limitation. As image databases become larger and larger, an image retrieval system needs to integrate all of these individual image presentation techniques so that images can be found both effectively and efficiently.

An example of such kind of integrated image retrieval system has been shown using Google image search application. It is shown that Google image search has integrated caption browsing, category browsing, hierarchical category browsing, QBE, and QBK into a single IR system. The success of Google's search engine shows that in terms of design a retrieval system, the simplest is the best. The chapter also demonstrates that powerful image visualization techniques such as a Google earth like image map can be utilized to enhance user experience on an IR system.

References

1. WordNet. https://wordnet.princeton.edu/. Accessed Feb 2019
2. Schöffmann K, Ahlström D (2011) Similarity-based visualization for image browsing revisited. In: IEEE international symposium on multimedia
3. Quadrianto N, Kersting K, Tuytelaars T, Buntine WL (2010) Beyond 2D-grids: a dependence maximization view on image browsing. In: Proceedings of the international conference on multimedia information retrieval, pp 339–348,
4. Wang C et al (2015) Similarity-based visualization of large image collections. Inf Vis 14 (3):183–203
5. Google.com. Accessed Oct 2018
6. Islam M (2009) SIRBOT—semantic image retrieval based on object translation. PhD thesis, Monash University
7. Zhang D, Islam M, Lu G (2013) Structural image retrieval using automatic image annotation and region based inverted file. J Vis Commun Image Represent 24(7):1087–1098

Appendix: Deriving the Conditional Probability of a Gaussian Process

Given a set of data $\mathbf{x}_1, \mathbf{x}_2, \ldots, \mathbf{x}_N$ from a certain class C, and each feature vector $\mathbf{x}_i$ is a data point in R^D: $\mathbf{x}_i = (x_{i1}, x_{i2}, \ldots, x_{iD})$. A matrix $\mathbf{X} = D \times N = (\mathbf{d}_1, \mathbf{d}_2, \ldots, \mathbf{d}_D)^{\mathrm{T}}$ can be created as following:

$$\mathbf{X} = \begin{bmatrix} x_{11}, & x_{21}, & \ldots, & x_{N1} \\ x_{12}, & x_{22}, & \ldots, & x_{N2} \\ & & \ldots & \\ x_{1D}, & x_{2D}, & \ldots, & x_{ND} \end{bmatrix} = \begin{bmatrix} \mathbf{d}_1 \\ \mathbf{d}_2 \\ \ldots \\ \mathbf{d}_D \end{bmatrix} \tag{A.1}$$

$\mathbf{X}$ is a Gaussian process and $\mathbf{X}$ follows a multivariate normal distribution: $\mathbf{X} \sim \mathcal{N}(\boldsymbol{\mu}, \Sigma)$, where $\boldsymbol{\mu}$ and Σ are the *mean* and *variance* of $\mathbf{X}$ which are determined by (A.2) and (A.3), respectively.

$$\boldsymbol{\mu} = \begin{bmatrix} \mu_1 \\ \mu_2 \\ \ldots \\ \mu_D \end{bmatrix} \tag{A.2}$$

$$\Sigma = \begin{bmatrix} cov(\mathbf{d}_1, \mathbf{d}_1), & cov(\mathbf{d}_1, \mathbf{d}_2), & \ldots, & cov(\mathbf{d}_1, \mathbf{d}_D) \\ cov(\mathbf{d}_2, \mathbf{d}_1), & cov(\mathbf{d}_2, \mathbf{d}_2), & \ldots, & cov(\mathbf{d}_2, \mathbf{d}_D) \\ & & \ldots & \\ cov(\mathbf{d}_D, \mathbf{d}_1), & cov(\mathbf{d}_D, \mathbf{d}_2), & \ldots, & cov(\mathbf{d}_D, \mathbf{d}_D) \end{bmatrix} \tag{A.3}$$

where $cov(\mathbf{d}_i, \mathbf{d}_j)$ is either a *covariance* or a *variance*:

$$cov(\mathbf{d}_i, \mathbf{d}_j) = E\left[(\mathbf{d}_i - \mathbf{u}_i)(\mathbf{d}_j - \mathbf{u}_j)^T\right] = E\left[\mathbf{d}_i \mathbf{d}_j^T\right] - \mathbf{u}_i \mathbf{u}_j^T \tag{A.4}$$

$$var(\mathbf{d}_i) = cov(\mathbf{d}_i, \mathbf{d}_i) \tag{A.5}$$

and E is the expected value.

© Springer Nature Switzerland AG 2019
D. Zhang, *Fundamentals of Image Data Mining*, Texts in Computer Science, https://doi.org/10.1007/978-3-030-17989-2

To predict a new data or a new set of data $\mathbf{X}_*$, $\mathbf{X}$ and $\mathbf{X}_*$ are concatenated and the concatenated data is a new GP which follows the following normal distribution:

$$\begin{pmatrix} \mathbf{X}_* \\ \mathbf{X} \end{pmatrix} \sim \mathcal{N}\left(\begin{pmatrix} \boldsymbol{\mu}_{\mathbf{X}_*} \\ \boldsymbol{\mu}_{\mathbf{X}} \end{pmatrix}, \begin{pmatrix} \Sigma_{\mathbf{X}_*\mathbf{X}_*}, & \Sigma_{\mathbf{X}_*\mathbf{X}} \\ \Sigma_{\mathbf{X}\mathbf{X}_*}, & \Sigma_{\mathbf{X}\mathbf{X}} \end{pmatrix} \right) \tag{A.6}$$

Then, the probability of the new data $\mathbf{X}_*$ given the observed data $\mathbf{X}$ is given by (A.7):

$$p(\mathbf{X}_*|\mathbf{X}) = \mathcal{N}\left(\boldsymbol{\mu}_{\mathbf{X}_*} + \Sigma_{\mathbf{X}_*\mathbf{X}}\Sigma_{\mathbf{X}\mathbf{X}}^{-1}(\mathbf{X} - \boldsymbol{\mu}_{\mathbf{X}}), \Sigma_{\mathbf{X}_*\mathbf{X}_*} - \Sigma_{\mathbf{X}_*\mathbf{X}}\Sigma_{\mathbf{X}\mathbf{X}}^{-1}\Sigma_{\mathbf{X}\mathbf{X}_*} \right) \tag{A.7}$$

To prove (A.7), let's first formulate (A.6) into a general case of two multivariate normal distributions. Suppose $\mathbf{X}_1$ and $\mathbf{X}_2$ are the two partitions of $\mathbf{X}$:

$$\mathbf{X} = \begin{bmatrix} \mathbf{X}_1 \\ \mathbf{X}_2 \end{bmatrix} \tag{A.8}$$

where $\mathbf{X} \sim \mathcal{N}(\boldsymbol{\mu}, \Sigma)$, $\mathbf{X}_1 \sim \mathcal{N}(\boldsymbol{\mu}_1, \Sigma_{11})$ and $\mathbf{X}_2 \sim \mathcal{N}(\boldsymbol{\mu}_2, \Sigma_{22})$.
Then, the following are the partitions of $\boldsymbol{\mu}$ and Σ, respectively:

$$\boldsymbol{\mu} = \begin{bmatrix} \boldsymbol{\mu}_1 \\ \boldsymbol{\mu}_2 \end{bmatrix} \tag{A.9}$$

$$\Sigma = \begin{bmatrix} \Sigma_{11} & \Sigma_{12} \\ \Sigma_{21} & \Sigma_{22} \end{bmatrix} \tag{A.10}$$

where Σ and Σ_{ij} are all the covariance matrices. Based on (A.4) and (A.5), it can be shown that the following properties of vector *variance* (*var*) and *covariance* (*cov*) are true [7, Chap. 7]:

(1) $var(\mathbf{AX} + \mathbf{b}) = \mathbf{A}\,var(\mathbf{X})\mathbf{A}^T$ (A.11)

(2) $var(\mathbf{X}_1 + \mathbf{X}_2) = var(\mathbf{X}_1) + var(\mathbf{X}_2) + cov(\mathbf{X}_1, \mathbf{X}_2) + cov(\mathbf{X}_2, \mathbf{X}_1)$ (A.12)

(3) $cov(\mathbf{AX}_1 + \mathbf{b}, \mathbf{X}_2) = \mathbf{A}\,cov(\mathbf{X}_1, \mathbf{X}_2)$ (A.13)

(4) $cov(\mathbf{AX}_1 + \mathbf{BX}_2, \mathbf{X}_2) = \mathbf{A}\,cov(\mathbf{X}_1, \mathbf{X}_2) + \mathbf{B}\,var(\mathbf{X}_2)$ (A.14)

(5) $cov(\mathbf{X}_1, \mathbf{X}_2) = cov(\mathbf{X}_2, \mathbf{X}_1)$ (A.15)

To compute the parameters of $p(\mathbf{X}_1|\mathbf{X}_2)$, we need to create an auxiliary random variable $\mathbf{Z}$ which is both the linear combination of $\mathbf{X}_1$ and $\mathbf{X}_2$ and *independent* of $\mathbf{X}_2$ [8, Chap. 7]:

$$\mathbf{Z} = \mathbf{X}_1 + \mathbf{A}\mathbf{X}_2 \qquad (A.16)$$

Then, the matrix $\mathbf{A}$ can be found by letting $cov(\mathbf{Z}, \mathbf{X}_2) = 0$:

$$\begin{aligned} cov(\mathbf{Z}, \mathbf{X}_2) &= cov(\mathbf{X}_1 + \mathbf{A}\mathbf{X}_2, \mathbf{X}_2) \\ &= cov(\mathbf{X}_1, \mathbf{X}_2) + \mathbf{A}var(\mathbf{X}_2) \\ &= \Sigma_{12} + \mathbf{A}\Sigma_{22} = 0 \end{aligned} \qquad (A.17)$$

which leads to

$$\mathbf{A} = -\Sigma_{12}\Sigma_{22}^{-1} \qquad (A.18)$$

and

$$E(\mathbf{Z}) = E(\mathbf{X}_1) + E(\mathbf{A}\mathbf{X}_2) = \boldsymbol{\mu}_1 + \mathbf{A}\boldsymbol{\mu}_2 \qquad (A.19)$$

Therefore, we have

$$\begin{aligned} E(\mathbf{X}_1|\mathbf{X}_2) &= E(\mathbf{Z} - \mathbf{A}\mathbf{X}_2|\mathbf{X}_2) \\ &= E(\mathbf{Z}|\mathbf{X}_2) - E(\mathbf{A}\mathbf{X}_2|\mathbf{X}_2) \\ &= E(\mathbf{Z}) - \mathbf{A}\mathbf{X}_2 \\ &= \boldsymbol{\mu}_1 + \mathbf{A}(\boldsymbol{\mu}_2 - \mathbf{X}_2) \\ &= \boldsymbol{\mu}_1 + \Sigma_{12}\Sigma_{22}^{-1}(\mathbf{X}_2 - \boldsymbol{\mu}_2) \end{aligned} \qquad (A.20)$$

where $E(\mathbf{A}\mathbf{X}_2|\mathbf{X}_2) = \mathbf{A}\mathbf{X}_2$ is due to both $\mathbf{A}$ and $\mathbf{X}_2$ are constants. $E(\mathbf{X}_1|\mathbf{X}_2)$ is the *mean* of $p(\mathbf{X}_1|\mathbf{X}_2)$.

By using (A.12) and (A.17), we have the following:

$$\begin{aligned} var(\mathbf{X}_1|\mathbf{X}_2) &= var(\mathbf{Z} - \mathbf{A}\mathbf{X}_2|\mathbf{X}_2) \\ &= var(\mathbf{Z}|\mathbf{X}_2) + var(\mathbf{A}\mathbf{X}_2|\mathbf{X}_2) + cov(\mathbf{Z}, -\mathbf{A}\mathbf{X}_2) + cov(-\mathbf{A}\mathbf{X}_2, \mathbf{Z}) \\ &= var(\mathbf{Z}|\mathbf{X}_2) + \mathbf{A}var(\mathbf{X}_2|\mathbf{X}_2) - \mathbf{A}cov(\mathbf{Z}, \mathbf{X}_2) - \mathbf{A}cov(\mathbf{X}_2, \mathbf{Z}) \\ &= var(\mathbf{Z}|\mathbf{X}_2) \\ &= var(\mathbf{Z}) \end{aligned}$$

$$(A.21)$$

Therefore, by combining (A.11)–(A.15), (A.18) and (A.21), we have:

$$
\begin{aligned}
var(\mathbf{X}_1|\mathbf{X}_2) &= var(\mathbf{Z})\\
&= var(\mathbf{X}_1 + \mathbf{A}\mathbf{X}_2)\\
&= var(\mathbf{X}_1) + \mathbf{A}var(\mathbf{X}_2)\mathbf{A}^T + \mathbf{A}cov(\mathbf{X}_1, \mathbf{X}_2) + \mathbf{A}cov(\mathbf{X}_2, \mathbf{X}_1) \quad (A.22)\\
&= \Sigma_{11} + \Sigma_{12}\Sigma_{22}^{-1}\Sigma_{22}\Sigma_{22}^{-1}\Sigma_{12} - 2\Sigma_{12}\Sigma_{22}^{-1}\Sigma_{21}\\
&= \Sigma_{11} - \Sigma_{12}\Sigma_{22}^{-1}\Sigma_{21}
\end{aligned}
$$

To summarize the above, we have obtained both the *mean* and *variance* of $p(\mathbf{X}_1|\mathbf{X}_2)$:

$$
\begin{aligned}
E(\mathbf{X}_1|\mathbf{X}_2) &= \boldsymbol{\mu}_1 + \Sigma_{12}\Sigma_{22}^{-1}(\mathbf{X}_2 - \boldsymbol{\mu}_2)\\
var(\mathbf{X}_1|\mathbf{X}_2) &= \Sigma_{11} - \Sigma_{12}\Sigma_{22}^{-1}\Sigma_{21}
\end{aligned}
$$

By substituting $\mathbf{X}_1$ and $\mathbf{X}_2$ with $\mathbf{X}_*$ and $\mathbf{X}$ respectively, (A.6) is proved.

An alternative proof of (A.7) is to use Bayesian theorem by working out the joint probability $p(\mathbf{X}_1, \mathbf{X}_2)$ and the marginal probability $p(\mathbf{X}_1)$ [9, Chap. 7].

$$
\begin{aligned}
p(\mathbf{X}_1, \mathbf{X}_2) &= \frac{1}{(2\pi)^{D/2}|\Sigma|^{1/2}}\exp\left\{-\frac{1}{2}\left[(\mathbf{X}_1 - \mu_1)^T, (\mathbf{X}_2 - \mu_2)^T\right]\begin{bmatrix}\Sigma_{11} & \Sigma_{12}\\ \Sigma_{21} & \Sigma_{22}\end{bmatrix}^{-1}\begin{bmatrix}\mathbf{X}_1 - \mu_1\\ \mathbf{X}_1 - \mu_1\end{bmatrix}\right\}\\
&= \frac{1}{(2\pi)^{\frac{m}{2}}|\Sigma_{11}|^{\frac{1}{2}}}\exp\left\{-\frac{1}{2}(\mathbf{X}_1 - \mu_1)^T\Sigma_{11}^{-1}(\mathbf{X}_1 - \mu_1)\right\}\\
&\quad \times \frac{1}{(2\pi)^{n/2}|\mathbf{A}|^{1/2}}\exp\left\{-\frac{1}{2}(\mathbf{X}_2 - \mathbf{b})^T\mathbf{A}^{-1}(\mathbf{X}_2 - \mathbf{b})\right\}\\
&= \mathcal{N}(\mathbf{X}_1, \boldsymbol{\mu}_1, \Sigma_{11}) \times \mathcal{N}(\mathbf{X}_2, \mathbf{b}, \mathbf{A})\\
&= p(\mathbf{X}_1) \times p(\mathbf{X}_2)
\end{aligned}
$$

$$(A.23)$$

where $m + n = D$ and

$$
\mathbf{A} = \Sigma_{22} - \Sigma_{21}\Sigma_{11}^{-1}\Sigma_{12} \qquad (A.24)
$$

$$
\mathbf{b} = \boldsymbol{\mu}_2 + \Sigma_{21}\Sigma_{11}^{-1}(\mathbf{X}_2 - \boldsymbol{\mu}_2) \qquad (A.25)
$$

Therefore, the conditional probability of $p(\mathbf{X}_2|\mathbf{X}_2)$ is given as follows:

$$
\begin{aligned}
p(\mathbf{X}_2|\mathbf{X}_1) &= \frac{p(\mathbf{X}_1, \mathbf{X}_2)}{p(\mathbf{X}_1)} \\
&= \mathcal{N}(\mathbf{X}_2, \mathbf{b}, \mathbf{A})
\end{aligned}
\tag{A.26}
$$

By substituting $\mathbf{X}_1$ and $\mathbf{X}_2$ with $\mathbf{X}$ and $\mathbf{X}_*$ respectively, (A.7) is also proved.

Index

© Springer Nature Switzerland AG 2019
D. Zhang, *Fundamentals of Image Data Mining*, Texts in Computer
Science, https://doi.org/10.1007/978-3-030-17989-2

Printed in the United States
By Bookmasters